The Hologram

The Hologram

Principles and Techniques

Martin J. Richardson
De Montfort University, Leicester, UK

John D. Wiltshire
Independent Consultant, UK

Registered Offices
John Wiley & Sons, Inc., 111 River Street, Hoboken, NJ 07030, USA
John Wiley & Sons Ltd, The Atrium, Southern Gate, Chichester, West Sussex, PO19 8SQ, UK

Editorial Office
The Atrium, Southern Gate, Chichester, West Sussex, PO19 8SQ, UK

For details of our global editorial offices, customer services, and more information about Wiley products visit us at www.wiley.com.

Wiley also publishes its books in a variety of electronic formats and by print-on-demand. Some content that appears in standard print versions of this book may not be available in other formats.

Library of Congress Cataloging-in-Publication data applied for

ISBN: 9781119088905

Cover design by Wiley
Cover image: © Visuals Unlimited, Inc./Carol & Mike Werner/Gettyimages

Set in 10/12pt Warnock by SPi Global, Pondicherry, India

10 9 8 7 6 5 4 3 2 1

Contents

Foreword

It was a great honour to receive the invitation to write this foreword. Holography has been part of my life both at the research level, applied to security of documents and products, and at the academic level, as a powerful tool to teach many of the complex aspects of optics.

The idea behind this book clearly detaches it from existing holography books that focus on the physics of the technique, requiring some considerable background in optics and mathematics, on the exquisite chemistry around the recording of the hologram, or on the artistic concept that surrounds this fabulous creative tool. In this book it is possible to navigate the interfaces between various types of knowledge involved and, when essential, the required physical/optical/chemical concepts are explained in a simple and pragmatic way, allowing the content to be easily explored by someone not entirely familiar with the subject, or to be appreciated by a specialist due to the simplified and abridged approach.

Reading this book reminded me of an anecdote (that I adapted to holography) about a complex holography camera that was having problems of consistency for several months. After using all the expertise available from scientists of all possible areas, the institution decided to call an old holographer who had worked in holography all his life. After a detailed analysis, the holographer fastened one screw with the proper torque and the holographic camera immediately started to give wonderful results. The institution was profoundly thankful to the holographer but considered the cost to be unexpectedly high for the activity performed. As a reply, the invoice from the holographer detailed: 1% of the cost – fastening the screw, 99% of the cost – knowing which screw to fasten!

This simple joke applies perfectly to the challenging complexity behind holography and makes us aware of something that is (apparently more than in other fields) funda-mental and, above all the academic/scientific knowledge, required to make good holo-grams: Experience. This book also conveys to the reader know-how gathered from several decades of experience, and this is undoubtedly a fundamental instrument if someone wants to take holography to the next level.

In conclusion, whether a researcher in a science institute, a teacher or a student in an optics class, or an artist in a holography studio, this book is a highly valuable tool for those starting to take the first steps on the difficult journey that is holography, and an excellent complement to the physics and chemistry books for those more advanced in the subject. Joyce Carol Oates, an American writer, once wrote that, "Beauty is a ques-tion of optics. All sight is illusion." If holography is an optical illusion, it is undoubtedly the most beautiful one.

Dr Alexandre Cabral

Preface

Hologram – The Thinking Picture

With the legacy of the inventions of Lippmann, Gabor *et al.*, we are lucky enough to have been the generation who have experienced the dawn of the *Age of Holography* first hand. Still, when we think of holograms, we think of the future, a place full of wondrous inventions. A place where driverless cars defy gravity, civilisation is established on Mars or cities are built under Earth's oceans. And then, of course, holographic images that materialise in thin air and communicate with each other using artificial intelligence, indistinguishable from human consciousness.

Such is the legacy of science fiction. In reality, this vision of the future may fall far beyond the laws of physics, but nevertheless there are some truly astounding developments taking place right now in the field of holography. I know, because as a research professor at De Montfort University, I've had the privilege of experiencing some of the world's most incredible three-dimensional holograms, and it's a glimpse into a future I want to share with you.

It could be that the laws of physics, outlined in Chapter 2 of this book, will prevent holography from fulfilling the science fiction vision of *Star Wars*. Instead, the holographic medium serves another function, a function that underlines the very nature of technological advance, as science strives to catch up with science fiction. The spin-offs are often more interesting than the original research intention! The fantasy of holography is a conceptual lubricant that facilitates the birth of other great ideas. Could it be, then, that the intrinsic, but unvoiced, value of holographic illusion points the way for next-generation immersive augmented reality, promising to evolve into something other, something unpredictable?

When Microsoft recently announced its technical breakthrough toward interactivity with holograms, it was a jaw-dropping moment for those within the holographic research community. The claim that *Windows* was about to enter our physical world through holographic technology owes much to the dream of science fiction and certainly adds another chapter to the history of three-dimensional imaging.

It took me several days of thought regarding the implications this would have on the research community and, after going through my initial feeling of elation, thoughts slowly slipped into its darker meaning. Was Microsoft misleading the public into thinking they had found the Holy Grail of 3D? The thought of holograms populating our everyday lives also felt somehow unsettling. It simply didn't align itself with current experience and, therefore, something seemed intrinsically wrong. Was the world on the brink of really merging with the digital Matrix? Was it a mirror rather than a window – a mirror reflecting another's identity, thoughts, desires and, therefore, needs? The idea of the mirror seen through another's eyes – someone else's view of reality – seemed beyond our perception, because Microsoft's *HoloLens*™ system threatens to invade the small amount of unencumbered reality we currently have, a space rapidly diminishing because of the ubiquitous screens on our walls, on our desks and in our pockets. *Real space is an endangered, diminishing asset.*

However, it quickly became clear that this interaction with our physical environment was through a head-set that superimposes the Microsoft operating system on the actual world. The Microsoft *HoloLens*™ system may not be holographic in the purest sense ("What is *Not* a Hologram?" Section 1.11 of this book), but the fact that Gabor's word "hologram'" continues to inspire innovation in the twenty-first century means our trip is far from over!

Perhaps The *HoloLens*™ could be said to be the modern-day equivalent to "Pepper's Ghost", a historical device used to create spectacular visual illusions by the use of projection systems, explained in Chapter 1. By utilising an angled, partially reflective surface, of which the viewer is not aware, between the audience and the main subject of a display, it is possible to produce a ghostly or ethereal image which appears to the audience to be superimposed in the same space as the principal ("real") display. Documentation of such a principle is recorded as far back as the sixteenth century in the writings of Giambattista della Porta. Later, such inventors as Rock [1] have attempted to improve the technique by suggesting improvements such as a method to hold a light, foil screen in position with a minimum of wrinkles in the film, which would normally be detrimental to the reflected image quality and thus to the illusion; ways to improve the brightness and contrast ratio of the projected images, and ways to eliminate extraneous light reflecting in the mirror screen, which tend to reduce the effectiveness of the illusion. Maas [2] also describes ways to improve the presentation of the basic principle. O'Connell [3] has shown ways of using this technique for video tele-presence methods.

In 1987, Stephen Benton, one of the world's great holographers and Professor at MIT, suggested that a stereographic three-dimensional hologram display should be confined within a limited viewing space (the Benton Alcove Hologram [4]) so as to restrict the viewer from coming into contact with the angular viewing limitations of the holographic image, which we all agree tend to be the "Achilles Heel" of display holography.

Recent technological developments by Zebra Inc. and XYZ have improved horizontal viewing angles and also provided vertical parallax by digital ray-tracing techniques, described in Chapter 8. The restriction of the ability to view the image from oblique angles is a severe disadvantage in comparison with Denisyuk holograms recently produced by Yves Gentet, Colour Holographic and Hans Bjelkhagen. But Denisyuk holograms of this type, whilst providing exceptionally realistic images (*approaching facsimiles of reality*), do not have the ability to represent computer-generated animated images, such as may be realised by digital techniques.

In the 1980s, holographer Peter Miller produced a two-colour reflection hologram with integrated sound system, featuring an image of a Barracuda Car Radio. Proximity switches behind the glass plate caused an audio effect to "change channel" when the viewer placed a finger in the *real image of the channel selector button*. This was a brilliant innovation which pre-dated the modern iterations of interactivity with a hologram!

In common with so many modern optical systems, a key component of the Microsoft *HoloLens*™ technique is a *holographic optical element* (HOE). These optical devices have a similar effect to a conventional glass optic but take the form of a thin film that is optically clear and has unique abilities in the manipulation of light. Previous to the HOE, non-holographic optical elements, made with a mechanical ruling device, were used in spectrophotometers, for example, as a dispersive grating to divide the spectrum from a white source into its separate colours, on an angular basis. An HOE is a convenient and relatively low-cost component with a highly efficient grating made by laser imaging of a photosensitive material. It is possible, using holographic methods, however, to produce more complex optics such as diffusers which control the exact direction of scattered light, or volume holograms such as head-up displays (HUDs) which direct reflected light of a narrow band of wavelengths in a specific (off-axis) direction; for a *holographic* mirror, the angle of incidence is not necessarily equal to the angle of reflection! HOEs to assist the collection of solar power will inevitably follow. So, positive commercial statistics predicting the future of holography seem relatively clear-cut.

In fact, modern holography offers many alternatives to light-shaping devices in industry and may be compared with the role electronic circuits and microprocessors played at the beginning of the 1960s as an alternative to the valve. As mass-produced holographic optical elements start replacing micro-lens arrays, and holographic phase memory is poised ready to replace today's standard magnetic hard drives, each has commercial potential previously thought impossible.

It remains to be seen if our ever-increasing dependence on technology will impair our physical or mental faculties and our adaptability to nature, but we do know that the advantage modern holography gains over existing technology will be long term and, in some cases, life-changing.

Creative computing will play a major role in the development of computer synthetic holograms within numerous applications including the arts, entertainment, games, mobile applications, multimedia, web design and other pervasive interactive systems. Due to the nature of these applications, computing technology needs to be developed specifically to tackle the conceptual complexity that does not exist in other applications. The challenges faced by creative computing come from the need to originate applications that involve knowledge in the disciplines of the Humanities and Arts more traditionally used to describe activities at human behaviour level. The main feature of these applications is that a creative system will directly serve people's needs to improve quality of life. It is the rapid development of computing technology that will enable new creative industry and it is also this rapid development that requires serious academic discussion.

Creative computing supports the vision that computing technology will become an integral part of the design industry, where computing offers new design tools for artists and designers to extend the traditional products, and computing technology itself will be developed and enriched by reference to knowledge from the Humanities and the Arts.

Today, holograms are standard security issue on bank cards and banknotes, event tickets, postage stamps and passports – all aimed specifically at halting counterfeiting. They are a typical component in the validation of safety-critical items – such as medicines and machine parts – and therefore save lives.

The list of applications of holography will increase in length as a growing number of five-star research labs in universities and technical companies, including Microsoft, find new applications for these amazing devices. We are developing new types of holograms with the long-term aim of progressing the medium beyond its ability simply to capture and replay three-dimensional images, pursuing their general ability to diffract and manipulate light. Extensive technical documentation concerning holography has established it as an exciting, emerging medium. However, its potential still remains relatively untapped. So, how did we arrive at this juncture of technology? Why does holography have the potential power to change the way we see our world and, as we start our journey into the *Age of Photonics*, where did the holographic journey begin?

The word "hologram" means many things to many people. The word was used by Professor Stephen Hawking as a metaphor to describe concepts in quantum mechanics. Hawking related the holographic principle with that of the need to explain the anomalous behaviour at a black hole, comparing the way it flattens time and space with the way a two-dimensional surface of a holographic recording carries a three-dimensional image. Other theoretical scientists suggest that the universe has qualities resembling a hologram, in that information about the whole exists in every constituent part. I'm reminded of William Blake's poem *Auguries of Innocence*:

> To see a world in a grain of sand
> And a Heaven in a wild flower
> Hold infinity in the palm of your hand
> And eternity in an hour

The word may also be used to describe the complete works of Shakespeare or a description of time, for example an organic life span from conception to death. Others tailor the word to promote idiosyncratic philosophy. The authors revert to the Greek roots of Gabor's term "holo" (entire) and "graph" (message).

In the following chapters, the intention is to offer a stepping stone for those who have an interest in this fascinating area of holographic imaging. We hope to provide an entry point into the philosophical and practical aspects of hologram-making; to understand how and why some of the holograms with which we are familiar today were made, and what the future holds for a relatively young technology as the related science develops.

Martin J. Richardson

Notes

1 James Rock. Patent application US20070201004A1: Projection apparatus and method for Pepper's Ghost illusion.
2 Uwe Maas – Musion Eyeliner 3D Projection www.eyeliner3d.com
3 O'Connell, I. (2009) "Video Conferencing Technique". *New Scientist*, 26 November.
4 Benton, S.A. (1987) "'Alcove' Holograms for Computer-Aided Design," *Proceedings of SPIE*, 0761, True Three-Dimensional Imaging Techniques and Display Technologies, 53.

Dedications and Acknowledgements

Martin J. Richardson

Dedicated to my daughters Elizabeth and Florence, who are inspirational, and my partner Nicky, for persevering more than thirty years of holographic mayhem, thank you!

John D. Wiltshire

For Carol, Jonathan and Darren.

Many thanks to Martin for the invitation to join him in the creation of this book. After 45 years working in the production, transport and recording of light, I hope my experience in practical issues will be useful to readers.

My uncle, Harold Swannell, an electrician at the Royal Small Arms Factory, Enfield, inspired and nurtured my lifelong interest in electricity, light and chemistry over sixty years ago.

Later, Joyce and Stanley Wiltshire, my late mother and father, coped patiently with explosions, fires and evil odours in our home throughout my trying youth. Thank you.

My Mum had ten siblings – her legacy: "The Turner Heritage."

Thanks to everyone who lived the "Holography Dream" together at Applied Holographics from 1983, during my 14 years at Braxted Park, and especially to Paul Dunn and Andrew Rowe for their recent help as I sought to remember and record those halcyon days.

My earliest inspiration for holography came from the late Graham Saxby, Nick Phillips and Steve Benton.

For the inspiring technical discussions I've had over the years with Peter Howard, Howard Buttery, Dave Oliff, Simon Brown, Jeff Blyth, Craig Newswanger, Peter Miller, David Winterbottom, Nigel Abraham, Mike Medora, Brian Holmes, Gideon Raeburn, Hans Bjelkhagen, Patrick Flynn, Satyamoorthy "Kabi" Kabilan, Jonathan Wiltshire and many other great scientists and engineers.

For my friends who didn't make it this far: Rob Rattray, Hamish Shearer, Micky Finlay and my soulmate Barney, I'm carrying the baton.

Thanks for irreplaceable contributions to this book and 20 years of friendship with Alexandre Cabral.

My sincere gratitude to project editor Samanaa Srinivas at John Wiley & Sons for invaluable help and advice in the realisation of a working manuscript.

To my friends throughout Europe – still love you – *back soon*!

Thank you.

About the Companion Website

Don't forget to visit the companion website for this book:

www.wiley.com/go/richardson/holograms

There you will find valuable material designed to enhance your learning, including:

- Video
- Figure PPTs

1

What is a Hologram?

1.1 Introduction

First, we should define in technical terms what precisely is meant by the term "hologram" and discuss some of the important milestones in the development of the technique.

Some years ago, holography was tauntingly dubbed "the solution looking for a problem" and it has taken some time to establish the true nature of the technology and disperse some of the urban myths which it seemed to attract so readily in earlier years.

We shall mention some of the important types of hologram recording which are possible, and end this chapter with a view of the public perception of holography; explaining a number of 3D systems which are frequently labelled "holograms" by the public, but which do not realistically meet the criteria for inclusion in this category.

Before embarking on a career in holography, remember that the holographer has to be prepared for the frequent request to re-create R2-D2's projection of Princess Leia, and also patiently to stand fast through the knowing wink accompanying the statement "Oh yes that's the method where the broken hologram still contains all of the image!"

1.2 Gabor's Invention of Holography

The word "hologram", coined by Denis Gabor from Greek roots, seems to be difficult to define precisely, so the authors prefer to assume the most appropriate English meaning to be "*the entire message*".

Of course, in an era where the Hellenic Institute of Holography plays a prominent role in development of new techniques for ultra-real representation of museum artefacts, the word has now ironically become "anglicised" to the extent that it now translates directly back into modern Greek as "*ολόγραμμα*"!

Gabor's intention was to refer, in the name, to the unique ability of this technique to record an incoming wave of light from an object in terms of both its phase and amplitude. Gabor was working in electron microscopy when he observed that interference recording was a way to achieve recordings of ultra-high resolution without the difficulties introduced in optical systems by the limitations of recording materials, lenses and conventional optics.

The Hologram: Principles and Techniques, First Edition. Martin J. Richardson and John D. Wiltshire.
© 2018 John Wiley and Sons Ltd. Published 2018 by John Wiley & Sons Ltd.
Companion website: www.wiley.com/go/richardson/holograms

The hologram differs greatly from the simple *photographic* recording of an image based solely on the amplitude of light arriving from the field of view of the optical system, because when we look at a black and white film exposed in an everyday camera, after applying developer, we see a perfectly recognisable "negative" image of the subject beyond the lens. In comparison, in its simplest form, the hologram recording itself, in the appropriate high-resolution silver halide film or plate, tends to be something that, at first sight, is a meaningless jumble of lines or zones in varying tones.

So, what is the "hologram"? In the etymological sense, in the early years there was a move to call the recording itself, the "holograph" and the reconstructed image, the "hologram". This seemed a logical and useful division, as it is often confusing during discussion in the workplace, as to whether we are referring to the glass plate in the laboratory or to the image which springs forth from it when the laser is switched on.

However, this terminology appears to have fallen by the wayside, and today's dictionary does not generally acknowledge the word "holograph" except in a separate sense with reference to the special legal value of handwritten documents.

When we look though a microscope at a silver halide holographic recording plate which has been exposed to a suitable "standing wave" of interference, and processed in a suitable developer and either "fixed" or "stopped", we see black and white lines in what appears to be a random pattern, or at least a pattern which does not appear to relate directly to the subject of the recording.

Its fringes might well, in some cases, be predominantly linear – leading to the concept of "surface-relief holograms". If the subject matter is relatively close to the plate, we may begin to recognise the shape of the object, but in a true "redundant" *Fraunhofer* hologram, it is quite impossible to relate the pattern to the subject matter (see Figure 1.1, which shows a magnified image of part of the recorded pattern in a holographic plate and, beside it, a black and white photograph of the recorded 3D image seen when the hologram was lit with helium–neon laser light).

So what is happening here?

The conditions for recording a hologram involve the use of a coherent light source. Lasers were not available in the era when Gabor invented holography, but nowadays we have a wide choice of laser types that can produce holograms, which will be detailed later.

Using a single laser, in the simplest format, we can arrange for one part of its emitted light to be incident upon a three-dimensional object. The remainder of the beam travels towards a high-resolution recording plate, and the light direct from the laser ("reference beam") coincides near the plate with light reflected from the 3D object ("object beam"). The light reflected from the object contains information about the shape and tonality of the object and is "coherent" with the light arriving at the recording plate direct from the laser (Figure 1.2). In the vicinity of the plate these two beams "interfere" to produce a "standing wave of interference" which can be recorded in the photosensitive emulsion on the plate provided certain conditions of stability exist – by definition, this "standing wave" *must not move or change* during the recording process.

So what is the nature of this "standing wave"?

We are familiar with the classical experimental demonstration by Thomas Young early in the nineteenth century. By using sunlight issuing from a tiny hole in a window blind (which was a simplistic way to provide a beam of partially coherent light), he was thus able to demonstrate the wave nature of light. This was achieved by passing light

(a) (b)

Figure 1.1 (a) The amplitude fringe recording; (b) the image it produces in laser light.

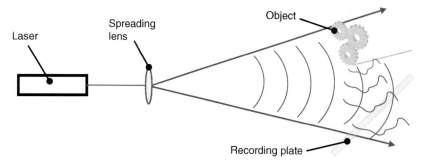

Figure 1.2 Making a transmission hologram with a single laser beam.

from this single source through two adjacent narrow slits in such a way that the two waves issuing from slightly displaced sources continued towards a screen. His famous sketch in Figure 1.3 was made by the visualisation of waves on water.

Now that we are routinely able to utilise laser light, we can easily demonstrate the analogous wave effect in electromagnetic radiation. If the screen CDEF is stationary, the extended row of spots which results demonstrates the interference of light from the two separate sources A and B.

As shown in Figure 1.3, the "wave fronts" from the two sources provide an orderly sequence of high and low intensity in accordance with the distance between the slits,

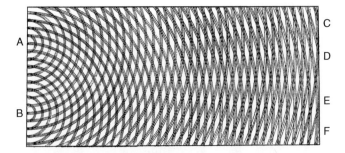

Figure 1.3 Young's slits visualisation [1].

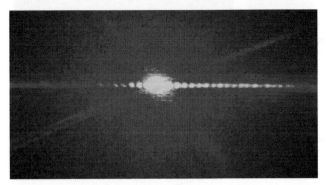

Figure 1.4 Twin-slit diffraction of green and red lasers.

the wavelength of the light and the distance of the screen from the slits (i.e. the angle between the beams).

Of course, if we were to place a sheet of photosensitive film in the position of the line marked CDEF by Young, in an optical set-up, we could record an interference pattern provided the "standing wave" was stationary.

Nowadays, we can very easily use a laser as a fully coherent light source and Figure 1.4 shows the effect of inter-changing the laser wavelength between recordings with the same slit apertures, in this case a pair of thin (0.1 mm) lines spaced by 1.5 mm etched into a black-developed photographic plate.

It is clear that increasing the distance between the slits in Figure 1.4 or reducing the distance to the screen will radically change the frequency of the fringes in accordance with "the grating equation" (see Chapter 2).

This fact will also lead us to see the advantage of the "off-axis" methods of recording holograms, later invented by Upatnieks and Leith in the USA simultaneously with the work of Denisyuk in the USSR; but it will also show the necessity for a spectacular improvement of the resolution characteristics of the films used for this more advanced form of holography. Recording materials for holography are discussed in detail in Chapter 5.

1.3   The Work of Lippmann

Gabriel Lippmann, working at the turn of the twentieth century, effectively anticipated the resolution and silver halide grain-size requirements of holographic recording materials. Using natural light, Lippmann's technique involved the production of colour photographs by a process which bears a close relationship to today's reflection holography. He was awarded the 1908 Nobel Prize for his work.

Essentially, he recognised the need for fine-grain silver halide emulsions and a method to process them which produced a transparent layer, rather than using the simple black metallic silver grains, which normally produce image contrast with the white background, that we associate with traditional "black and white" photography.

Lippmann produced photo emulsions with grain size of the order of 50 nm. This is difficult to do, even nowadays, for reasons explained later in Chapter 5.

By sandwiching this emulsion in intimate contact with a mirror surface in the form of liquid mercury, Lippmann was able to set up a recording process whereby a standing wave was produced by light reflected from the mirror, as it met light entering from the lens of his camera. If the camera was entirely still, the stationary wave formed by each individual wavelength of light would be recorded in the appropriate position in the photograph.

But, in common with today's "full-colour holography", achieved by coherent laser light, the ability of the emulsion to record a range of wavelengths simultaneously in a single zone of the recording material was limited. Furthermore, the use of incoherent light to produce a planar grating in the depth of the emulsion effectively meant that Lippmann was limited to a very thin layer of his scattering emulsion.

The natural "speed" of a photographic material tends to be proportional to its grain size, so Lippmann's exposure times were long – requiring great stability of the equipment and the subject itself; this still presents a problem today in modern holography. Processing fine-grain emulsion is also complicated, as explained in Chapter 6; these minute crystals of silver halide may well be soluble in chemicals designed for use with more typical, everyday photographic materials.

We will return to the connections between Lippmann's work and that of Denisyuk over 60 years later, and to the complex issues of photographic emulsion-making in subsequent chapters.

1.4   Amplitude and Phase Holograms

The first holograms were, quite naturally, recorded in silver halide material. Given its natural speed, this has obviously been the material of choice for the photographic industry for many years, only recently challenged by digital electronic technology. Only the problems outlined above, as addressed by Lippmann, regarding the resolution of the recording materials for holographic use have prevented silver from remaining the material of choice to this day, when new materials have emerged. We discuss the alternative recording materials for holography in Chapter 5.

Silver halide, as explained in detail in Chapter 5, produces, after initial development, grains of black silver metal (which may appear tinted dark red or green in transmitted light, as we will explain later) that produce an "amplitude" record of the zones/fringes which are associated with areas of high intensity (additive or constructive) interference of the light in the standing wave recorded, and which, in turn, represent the interaction of the reference beam with the object wave.

This record is called an *amplitude hologram*: a three-dimensional matrix of zones of clear gelatin interrupted by grains of opaque silver metal which prevent the direct passage of light rays through the layer. This works relatively well for transmission holograms, where, as we will show later, the fringe microstructure is principally perpendicular to the layer; a predominantly superficial recording.

But if we then continue the emulsion processing with a bleaching solution, we can change the amplitude hologram into a phase-modulated microstructure. The black grains of silver which prevent the passage of light rays through the layer can be changed advantageously into translucent crystals of silver bromide. This is a very simple chemical step:

$$Ag - e = Ag^+$$

$$Ag^+ + Br^- = AgBr_{crystal}$$

The loss of an electron is a simple oxidation process which can be brought about by any number of common oxidising agents (for example, a solution of ferric [iron III] sulphate). Loss of the electron results in the creation of a silver ion, Ag^+, which then combines in solution with a negative bromide ion in the bleaching solution to produce translucent (pale yellow) silver bromide.

Now that we have effectively removed attenuating black material from the layer, we have a condition where light rays entering the layer are not significantly absorbed, but are now subject to diffraction at interfaces within the layer which are manifested by local "image-wise" modulation of the refractive index of the constituents of the layer.

Now, whereas silver halide is historically the most important recording material, we must consider other, later materials which directly provide phase modulation. These include dichromated gelatin (DCG), which found an important commercial outlet in hologram pendants for jewellery, and, at the opposite end of the market, the use of holograms as optical elements such as the head-up displays which were introduced in fighter aircraft in the 1970s.

The DCG medium became unfashionable later due to technical difficulties, general chemical undesirability and archival problems, and has been replaced predominantly by modern photopolymer materials in the role of direct phase-modulated materials.

1.5   Transmission Holograms

Simplistically, a transmission hologram is a recording where, *at the reconstruction or viewing stage,* the illumination source directs its light through the recording film or plate so that the viewer will interrogate the image from the opposite side to the light source, as seen in Figure 1.5.

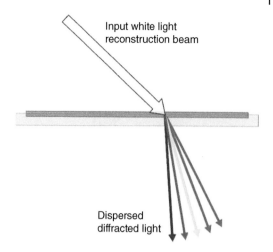

Figure 1.5 The dispersion of white light by a transmission hologram. Figure courtesy of Alex Cabral.

Input white light reconstruction beam

Dispersed diffracted light

This phenomenon is a function of the recording arrangement. Of course, the irony is immediately apparent that, as we discuss in detail in Chapter 8, the format for *recording* a transmission hologram, which is *viewed* from the side opposite to the illumination source, to enable the appropriate fringe structure to form, is a configuration where both object and reference beams arrive from the same side of the plate (the opposite is true for reflection holograms!).

The fact is that, despite a certain level of popular disbelief, a transmission hologram may be either a thin hologram or a thick (volume) hologram. This is entirely dependent upon the recording medium and we will detail in Chapter 5 why certain materials are more appropriate than others in the various hologram recording configurations, due to their ability to record either *surface* or *volume* microstructures.

But, as shown in Figure 1.5, the diffractive properties of a transmission hologram work in such a way as to disperse incident white light into its component colours. The hologram acts, in its effect, in a similar way to a prism, although the colour distribution appears reversed. This fundamental property of the transmission hologram made it impossible in the early years to view the image in white light, until the incredible invention by Dr Stephen Benton in 1967 of the *rainbow hologram*, which we describe later in detail and which has facilitated the major commercial outlet for the technology in security applications.

So, the embossed holograms with which we are so familiar are, in fact, transmission holograms. By mounting the transmissive layer upon a metallic foil substrate, we are able to view the image from the same side as the illumination source. Similar means have been used to display large-format transmission rainbow holograms; that is, by mounting a glass or film hologram layer upon a mirror before framing the whole assembly.

1.6 Reflection Holograms

In the simplest terms, reflection holograms are those which are illuminated *in the recon-struction or replay step* from the same side of the plate or film as the viewer sees the image. These holograms are often called *volume reflection* or *Lippmann* holograms.

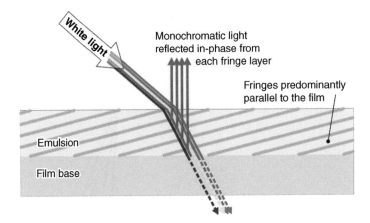

Figure 1.6 The planar index modulations (fringes) reflect light of a single wavelength only. Figure courtesy of Alex Cabral.

Note: The use of the term "volume hologram" works very well in the *technical* environment, but the authors have frequently been seriously frustrated in commercial discussions by the interpretation of this term as relating either to the quantity of holograms to be produced or to the fact that the image itself has depth or *volume*!

As with transmission hologram technology, the configuration for recording such a microstructure is the opposite of the viewing configuration, so that to create the required microstructure here, we must arrange for the recording *object* and *reference* beams to be incident from opposite sides of the recording medium. The result is that the fringes are typically planar within the layer and, like Lippmann's photographs, they tend to act somewhat like a mirror, rather than showing the prism-like dispersive effect of the transmission hologram. This is why the concept has been summarised as being "like a mirror with a memory".

This "mirror" has the quality of being monochromatic, so it acts rather like a filter to incident white light, in that a single wavelength is selectively reflected, whereas rays of all other wavelengths pass directly through the layer. This is because rays of light, shown in orange in Figure 1.6, whose wavelengths coincide with the particular selected frequency of the planar grating, constructively interfere, in accordance with Bragg's Law, at the successive parallel index interfaces, in order to create a reflected summation of their energy, whereas other wavelengths fail to meet the conditions for constructive interference. It is important to recognise that *Bragg diffraction* follows a mechanism of constructive interference, whilst thin holograms actually function quite differently.

Because the reflection hologram is thus able to reflect selected light of a certain wavelength, and of course this may be in the form either of a (simple) plane wave or a complex object wave, as previously discussed, the device itself may take the form of a clear, almost colourless layer, which bears a three-dimensional or animated image, so that such an extremely simple, compact, ethereal film layer is a very attractive proposition in security applications, where it may be used to overlay conventional printed graphics, allowing an unobstructed view of printed information on a document, in the "off-Bragg" conditions of illumination.

In this configuration, we see the vital importance of the phase (bleached) hologram set-up. Interestingly, if we make an exposure in the reflection mode to a suitable silver halide film, we can develop the film to a low density and if, at that time, we dry the film and view the hologram with strong illumination, we will see a very dim monochromatic reflection image in the black silver layer. If we imagine the fringes of this "Lippmann–Bragg" hologram to be of similar planar structure to the pages of a book, albeit tilted perhaps at a small angle to the surface of the film, as shown in Figure 1.6, then, of course, a ray of light entering the layer from the surface of the film is gradually absorbed by the layers of black metallic silver as it advances into the hologram layer. What light is reflected at the interfaces of gelatin and silver metal provides us with a dim, monochromatic image of our chosen subject.

As previously described, we can convert the metallic silver into translucent crystals of silver bromide, which incidentally has an unusually high refractive index (~2.23), the importance of which will become clear. Now, the layer appears clear and light is able to pass through the whole layer with little attenuation. More importantly, the emulsion layer comprises alternating planar zones of higher and lower refractive index in accordance with the modulated concentration of silver bromide in these individual planes. We have thus produced an *index-modulated* microstructure. The image of this phase hologram may be extremely bright and such a grating may be close to 100% efficiency in terms of its ability to diffract light *of a single specific wavelength*.

1.7   Edge-lit Holograms

Stretching back to the era of Leith and Upatnieks, there is a very interesting "special case" where the structure of a holographic grating falls between the "transmission" and "reflection" modes. This phenomenon has attracted sporadic research interest including some work by the true doyens of holography such as Leith, Phillips and Benton.

Due to the fact that the volume microstructure is basically mid-way between the defined "transmission" and "reflection" configurations, we have a hologram which, upon receiving illumination of an appropriate wavelength at an oblique angle within the support layer, as shown in Figure 1.7, is able to direct that light in such a way as to exit

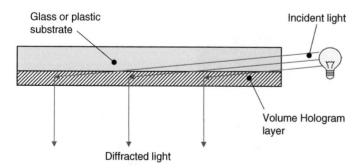

Figure 1.7 Light introduced into the edge of the substrate layer is diffracted so as to emerge from the surface of the assembly.

the film through its front surface in the form of a plane wave; or indeed in the form of a complex object wave with the ability to display a 3D or animated image.

The technique has been slow to deliver results of similar quality to the highly accomplished achievements in the conventional display and security arenas. At first sight, physicists are frequently excited by the apparent reassurance that the light which is successfully launched into the substrate is constrained therein by total internal reflection (TIR) at the front surface, but the holographer has to contend with the fact that, even in the event that index-matching surface couplers in the form of adhesives can ensure the passage of light from the illuminated substrate into the volume hologram itself, its angle of incidence upon the microstructure itself is critical and must comply with the Bragg condition. The authors have reason to believe that new recording materials, lasers and optical configurations may well lead this technology into widespread commercial use in the future.

1.8   "Fresnel" and "Fraunhofer" Holograms

The generic terms "Fresnel" and "Fraunhofer" holograms represent the extreme cases that we meet on the optical table when producing actual holograms. It is useful to define the meanings here because we will see later that the proximity of the "object source" of a hologram is a relevant factor in the way we record images in master and copy holograms, and certainly influences the limits of diffraction efficiency which might be expected from any recording configuration.

Fraunhofer was a German physicist active at the turn of the eighteenth century. His observation that a distant object obstructing a beam of light will produce a shadow whose periphery is characterised by surrounding lines of alternating light and dark, is an effect which becomes all too familiar when we begin to utilise lasers as a light source, and one which provides significant problems when we are masking down laser beams on the optical table or creating apertures to permit the passage of only the central portion of a laser beam into an adjacent part of the optical system.

Figure 1.8 is a photograph of a collimated laser beam falling on a distant screen. The edges of the aperture manifest themselves as a periodic sequence of lines of alternating

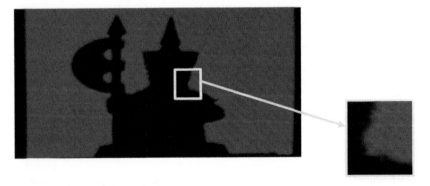

Figure 1.8 Sharp edges of the shadow of the model are contaminated by the effects of edge diffraction.

high and low brightness. Of course, this effect is not limited to the edges of an aperture but equally, or more importantly, affects the edges of every detail of the graphic image. So, for example, Gabor's slide subject which contained a text tribute including the names Fresnel, Young and Huygens, was rear illuminated such that the "edge diffraction" in the image seen on a screen beyond the slide was included in the light arriving at the recording material.

In the modern era we take it for granted that the "reference beam" will be launched from a slightly displaced position, such that it does not contain the shadow of the subject graphics.

However, looking in detail at the shadow of the model cast on a white screen by a collimated laser beam in Figure 1.8, we see the effect of edge diffraction at the periphery of the shadow, as well as the two edges of the aperture frame.

Wonderfully, unlike the simple photographic image of Figure 1.8 recorded at the recording plate where, unlike the holographic parallel, no coherent reference beam was present, a holographic recording at the same position can cope perfectly with this situation, where incident rays include edge diffraction and all other "spurious" features, and the fidelity of the hologram will thus benefit, in its ability to reconstruct the original scene, from *every* ray of light associated with the subject matter which arrives at the plate! The illumination of the recorded *holograph*, with the identical conditions of incidence to the original reference beam parameters (conjugate reference), will re-create the original wave front, with the effect that the image seen, unlike this *photograph*, will be a precise, sharp likeness of the subject of the recording at the precise position in which it was originally placed.

The Fraunhofer hologram is one where the object is distant from the recording medium; but to make such a statement of relative scale we need to qualify the size of the object itself. Fraunhofer's set-up is not unlike the method which Denis Gabor used in the original demonstrations of holography. In his case, Gabor found that the presence of the conjugate real image was particularly disruptive to the view of the subject.

The "off-axis" regimes of Denisyuk and Leith and Upatnieks were later to solve that problem in a single step. The term "redundant" may also be used to describe a hologram where the object is distant from the recording medium. This definition will also allow for the oft-quoted phenomenon where a shard of a broken glass hologram contains information about the whole subject matter. It is possible to photograph an image of the whole subject of the original hologram, even if only a small glass segment is available. The photograph in Figure 1.1(b) is a view through a small part of the "redundant H1 master" pulse laser hologram, shown in Figure 1.1(a), made by the authors of the singer Billie Piper. The camera lens is close to the master, looking through the plate, whilst it is illuminated by laser light. Of course, any part of the master hologram will provide a view of the whole portrait, so it is true to say that if the glass master was broken, each part would contain a worthwhile image of the subject. The unspoken "lost information" in this case is parallax, of course – the solid angle through which the object can be seen is clearly limited by the size of the shard, in a similar way as it was originally limited to a lesser extent by the size of the whole "un-broken" recording plate; after breaking, the window of view is far smaller, but still shows a *narrower* view of the *whole* subject.

A Fresnel hologram is the opposite case, where the diffuse subject is close to the recording material. Augustin-Jean Fresnel was a French physicist at the turn of the nineteenth century who was an important contributor to the wave theory of light.

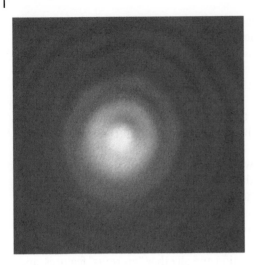

Figure 1.9 Circular wave front emanating from a point source.

Now, in holography, every individual point on the surface of the laser-illuminated object is effectively the source of a secondary wave. A segment of every wave will arrive at the hologram recording plate in accordance with the proximity of the diffuse object. In the case that the distance is short, the wave will intersect the recording material so as to produce a wide range of angles relative to the reference beam. The emission wave of each individual point on the surface of a diffuse object incident upon the recording material thus resembles the familiar form of a Fresnel "zone plate" (Figure 1.9).

The intersection of multiple circular wave fronts, as seen in Figure 1.9, emanating from every point on the surface of an object would lead to incredible complexity in the combined two-dimensional intensity profile. We can imagine, therefore, that the combined (volume) interference pattern, which is the summation of all such points on the diffuse surface, is an incredibly complex three-dimensional pattern, and as such this hologram recording will demand a whole new level of resolution (measured as a modulation transfer function, MTF) from the recording material as compared to the relatively simple, predominantly two-dimensional structure of the Fraunhofer hologram.

In the Fresnel hologram, as was mentioned previously, it is possible, to some extent, to recognise the object matter of the hologram; this "photographic" effect may well prevent us from achieving maximal diffractive efficiency and aesthetic appearance in the hologram; and its manifestation in the film or plate surface is often referred to by holographers as "burn out", as its amplitude tends to exhaust the dynamic range of the recording material locally and is often deleterious to holographic images.

1.9 Display Holograms

There are a number of holograms which have become icons of the technology. Exhibitions of display holograms have been held in every part of the world. In general, framed exhibition holograms are glass plate recordings which are second generation copies of original master recordings. Clearly, framed glass reflection holograms lend themselves particularly well to exhibition, as they can be lit conveniently by spotlights

Figure 1.10 Second generation transfer hologram.

attached to the ceiling (there is no reason why they cannot be lit from the floor level; this is known as "bottom reference").

As will be discussed in Chapter 8, master holograms can be transferred in their second generation into either rainbow transmission or reflection formats.

In Figure 1.10, my co-author's hologram, "The Art of Science" a 50 cm x 60 cm pulse-mastered glass display, has been produced in both formats. It is possible, in the case of a transmission rainbow hologram, to use one of two techniques:

1) To organise for the plate to hang in such a way that it can be lit from the reverse side. For example, the framed hologram can be elegantly suspended from the ceiling with wires.
2) To mount the hologram on a mirror such that it is treated as a "reflection hologram". This does, however, require that the processing of the plate uses a technique that provides a clear and predominantly colourless layer. It also introduces the possibility that the hologram and mirror assembly may require index matching to avoid "Newton's rings" (wood grain interference).

In more recent years, the digital holograms made by companies such as Zebra and XYZ Inc. have introduced new technology in which the image comes about by quite different means, and we shall describe this technology later. These techniques lend themselves to "tiling", where a large image can be assembled from a number of smaller component plates. The difficulty of producing large-format holograms is considerable; generally, 50 cm x 60 cm has become the observed size limit. Formerly, in the 1980s, Agfa Gevaert produced one-metre-square recording plates on 6 mm glass, and also supplied rolls of film on a triacetate base of one metre roll width. Of course, handling such large and heavy recording material presented significant difficulties in the holography studio and the chemistry laboratory!

For the purposes of our intention in this chapter to define the conventional techniques of holography and its associated terminology, it is appropriate to introduce everyday terms used in conventional holography. It is widely accepted that we refer to a first generation recording of subject matter or artwork as the "H1" and further image transfers as "H2" and "H3". (Only counterfeiters need concern themselves with H4!)

(a)

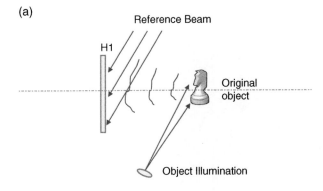

(b)

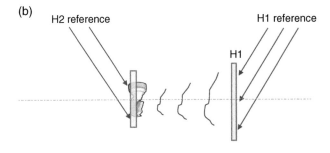

(c)

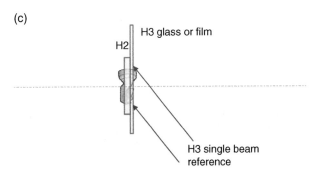

Figure 1.11 (a) First generation (H1) recording of a real object; (b) second generation (reflection) "image-planed" recording (H2); (c) the "contact copy" configuration for mass production.

The basic configuration of these optical arrangements is shown schematically in Figure 1.11. To summarise the typical procedure, the object or artwork (a three-dimensional item or model or a graphic display) is illuminated with laser light, and the coherent reference beam arrives simultaneously to produce a standing wave of interference, which is thus recorded in the "H1 transmission master", shown in Figure 1.11(a).

This plate is regarded as a permanent archival record of the subject matter. In Figure 1.11(b), this "H1" plate, with its "redundant" image displaced far from the surface, is illuminated by a laser (preferably of the same wavelength as the originating laser) from the opposite surface of the plate, to produce a real image of the subject, displaced from the surface. (The real image has the property of "pseudoscopia".)

A second, unexposed recording plate is then placed a similar distance from the H1 as was the original object, and a new coherent reference beam is introduced to record a new image of the original "H1" recording, as a second generation hologram; hence its title, "H2". In the case where the second reference beam is incident from the side opposite to the H1 master, as shown in Figure 1.11(b), a *reflection* hologram will result. If, however, the second reference beam is directed from the same side of the plate, a second generation (H2) *transmission* hologram is recorded.

This "H2" usually has the quality of being "image-planed." We generally now select the position of the subject of the hologram to be astride the recording medium to achieve the mysterious and exciting illusion for the viewer where the 3D subject appears to be situated partly in front of the display and partly behind the film or plate. In Chapter 10, we will refer to our hologram "Chelsea Library", where the image has a depth of more than two metres and intersects the film just beyond its mid-point, with spectacular effect. In my time at Applied Holographics, James Copp produced an incredible image of branches from a tree whose H2 transfer allowed an incredible real image projection when illuminated by laser. It was possible when viewing the hologram to walk into the scene, with the surreal effect that that twigs in the extreme projection appeared to project literally *into* the viewer's eye (without pain!).

In certain applications, for example for mass production purposes, the "H2" hologram, which, as previously suggested, could be either a glass or film layer, may be used as a second "master" hologram, as shown in Figure 1.11(c).

It is possible to use this hologram as the subject for contact copying. Film is typically laid upon the "H2" in order to allow a single laser beam to transfer the image from the "H2" into the new film layer, to produce the third generation "H3" image. This effect is explained more fully in Chapter 8.

1.10   Security Holograms

Holography has found its primary commercial application in the security industry. The general observation that any technique is frustratingly difficult to perfect means that an opportunity exists to use the product as a badge to mark a genuine product, provided a significant barrier is presented to those who may wish to attempt to counterfeit the device.

Holography offers such a barrier, not least because during the route to a mass-produced product there is a plethora of opportunities for error; only the most experienced holographer will avoid such pitfalls.

Applied Holographics was formed by Larry Daniels and Hamish Shearer in 1983 in order to exploit the use of reflection holography in display applications, but there was a rapid change of emphasis as soon as the potential as a security device was recognised by the security industry. The roots of that company in reflection holography were necessarily adjusted in favour of embossed holography after ABNH released the VISA dove in 1984.

Applied had, by that time, successfully commissioned its ruby pulse replicator "The HoloCopier" based upon fine-grain film produced by Ilford Ltd. Commercially, however, there was no option but to concentrate on the embossing method, in the light of the cost advantages of thermoplastic foils over silver halide, and also bearing in mind the existing compatibility between the printing industry with existing "hot-foil" products, which, in turn, offered a link to surface-relief embossing. Actually, that link was made rather less simple by the relatively small margin of temperature between the embossing and "hot-stamping" processes.

Embossed holography has been in the vanguard of security printing since that date and a large industry has developed. In recent years there has been a serious increase in the incidence of counterfeiting and it was suggested by David Pizzanelli for Smithers-PIRA [2] in 2009 that volume reflection holograms in photopolymer film would be the leading disruptive technology of the present decade (the exponents are running out of time!).

Embossing is a process which utilises surface-relief diffractive images to be pressed at very high speed on a roller system into cheap thermoplastic foil. Because the result is a definitively thin diffractive microstructure (transmission hologram), it disperses incident light into its individual colour components. Utilising Benton's "rainbow" technique, holographic images reflected from the silver foil substrate appear incredibly bright in aesthetically attractive, iridescent colours. The technique is capable of producing very large quantities of low-cost foil holograms, and for this reason, the security industry has used embossed holograms in the form of adhesive labels and hot foil for over thirty years, with English banknotes bearing characteristic holograms made by De La Rue Holographics since the millennium. Only now is counterfeiting becoming a serious problem for embossed holography, and the security industry is calling for new solutions, which might well involve full-colour reflection holograms, as previously suggested by PIRA.

The established methods of applying hot foil to banknotes, credit cards and packaging will need to be replaced with wholly new techniques for secure application of reflection holograms. The introduction of photopolymer as a recording medium for volume holograms has tended to displace from favour the idea of silver halide mass production pioneered by Applied Holographics in the 1980s.

At that time, Du Pont were discussing such a photopolymer material with AH, and Du Pont Authentication Systems Inc. went on to produce the material "Izon", which has famously protected products such as Nokia batteries and AMD hardware from counterfeit. Bayer Materials Science (now Covestro) has introduced its "Bayfol HX" material, which is a polychromatic recording film capable of high diffraction efficiency in a full range of colours, by the use of red, green and blue lasers to produce volume reflection holograms.

As we shall see in Chapter 5, one of the inherent problems of mass producing reflection holograms is the need to match the remarkable production speed at which embossed holograms can be pressed into thermoplastic foil. However, new developments, especially in laser technology, may well progress reflection holography into a new era of commercial usefulness.

1.11 What is *Not* a Hologram?

One of the frustrations commonly expressed by holographers is the continual receipt of enquiries that refer to techniques which cannot be considered to be holograms. 3D imaging is certainly not limited to holography; in the same way that holography is not necessarily limited to 3D imaging.

During the era of the original release of George Lucas's epic *Star Wars* films, as previously mentioned, the authors recall that holographers were inundated with requests to duplicate the effect shown in the scene where robot R2-D2 projected a 3D animated image of Leia (with sound!). But although labelled "a hologram", this imaginary technique does not fit easily into the scientific definition of Gabor's method of 3D imaging.

So, for the avoidance of doubt, we would like to address a number of categories of 3D imaging which are regularly referred to as holograms but which we feel do not necessarily fall in line with Gabor's invention to any great extent; and in some cases, not at all!

1.11.1   Dot-matrix Holograms

Clearly there is an open-ended zone of technical debate as to what constitutes a hologram. A prime example arose automatically within the industry in the 1980s when techniques such as "dot-matrix holography" were added to the portfolio of security hologram techniques. A dot-matrix "hologram" is a digitally originated array of dots wherein the individual units of exposure comprise an x–y matrix of plane gratings with various directional and colour qualities (Figure 1.12).

Such an array, although originally produced with lasers at Applied Holographics, could technically equally comprise an array of mechanically ruled gratings, and in the case of the *Kinegram*, originally produced by Landis and Gyr, similar effects are created by other mechanical means; or, in the case of CSIRO, by *electron beam lithography*. But because these techniques became involved with existing embossed holography work, and in the case of Applied Holographics were frequently combined as a mixed technique with "classical holography", they have been labelled "embossed holograms" with universal approval.

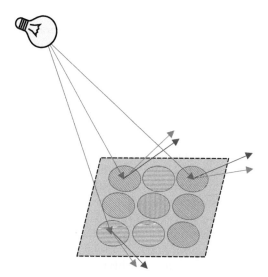

Figure 1.12 Reconstruction of a dot-matrix hologram.

1.11.2 Other Digital Image Types

In general, it is true there is something of a conflict between the terms "digital" and "hologram" in its fundamental sense (the entire message). Later, we discuss the techniques of digital reflection holography used by Zebra Inc. These holograms, in common with a similar technique later shown by XYZ Inc., appear to have the advantageous capability, since the surface matrix is divided into small pixels which are individually written, to produce hologram panels which could be aggregated together after exposure to produce very large 3D displays.

In common with the "one-step" methods invented by Haines for embossed holography, these "holograms" are essentially an array of pixels whose individual properties are predicted by ray-tracing techniques so as to enable each pixel to provide to the viewer the expected view of a large 3D subject when the viewer's eye is looking directly at that particular point in the surface matrix, from a certain viewing position. For these reasons, a pixel at the edge of an adjoining panel can be organised to coordinate spatially with the adjacent pixel in the next panel of a "tiled" array, thus inviting the possibility to build a matrix of panels capable of displaying a very large aggregated 3D image of an object.

1.11.3 Holographic Optical Element (HOE)

A holographic optical element is an optical device which has a similar effect to a conventional optic but takes the form of a thin film. We refer to a "*holographic*" element simply because we use the materials and techniques of conventional image holography to create the product. For example, a dispersive grating as used in a spectrophotometer to divide light from a white source into its separate colour components can be made by a mechanical ruling device. It is convenient and very cheap to produce a highly efficient linear grating by laser imaging on a photosensitive material and the manufacturer then also has the opportunity to incorporate the hi-tech "*holography*" buzz-word in the specification.

Whereas the above-mentioned techniques, when they apply to plane mirror optics, do not appear to fall in line closely with Gabor's original concept of the word "hologram", these are all important technical developments which have basically arisen because technologists have exploited his invention and, of course, they all have the common theme that their mode of operation can be defined as "diffraction".

However, there are techniques for 3D imaging which are far more difficult to classify as holograms. But the word "hologram" has captured the imagination of the public to such an extent that we have to accept that the following techniques are all frequently labelled as such!

1.11.4 Pepper's Ghost

The holographer has to endure the misplaced universal admiration of friends, upon their return from a USA vacation, for the "holograms" in Disneyland. In the main, these "holograms" are a variation on the theme of a well-known technique in the field of exhibition displays and theatre, which is commonly known as "Pepper's Ghost". Documentation of such principles is recorded as far back as the sixteenth century in the writings of Giambattista della Porta, an Italian scholar whose most important publication was entitled *Natural Magic*. But John Henry Pepper, a lecturer at the Royal Institution in the nineteenth century, gave his name to the modern iteration of the method.

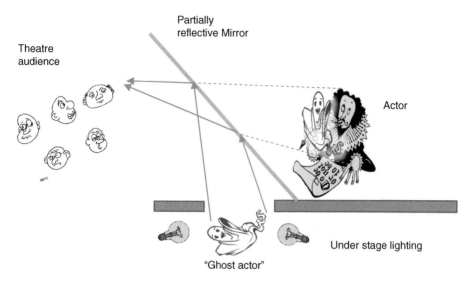

Figure 1.13 Schematic of the original theatre manifestation of "Pepper's Ghost".

Ironically, it was his colleague, Henry Dircks, who invented a specific display technique which he found it impossible to implement in theatre, but Pepper was able to modify the idea in order to make it feasible to apply the technique widely. In accordance with its title, the method was originally used to create the illusion of a ghost on stage.

An actor dressed in ghostly attire was hidden from the audience, brightly lit, in a confined area below the front stage level (in the orchestra pit). His reflected virtual image, on an angled glass plate, which filled the full aperture of the stage, its edges concealed on all sides, and whose presence was thus not perceived by the audience, appeared behind the glass plate, equidistant with respect to the actor's proximity to the glass, as shown in Figure 1.13. Thus, a three-dimensional virtual image of the "ghost" was apparently cast onto the stage area; an ethereal, floating, animated image through which the on-stage actor was able to walk!

There were problems, of course, with perspective effects, keystone distortion, image inversion, etc. and Pepper's system, for example, included a first mirror to aid with such difficulties.

Recently, this method has received excellent publicity from London Theatre and even TV news coverage of the world's first "Hologram Protest March" in Madrid. TV, newspaper and internet coverage of this event explained that the Spanish Civil Rights cause was advanced by use of a projection of this type featuring marching protestors. Involvement in a live demonstration for such a cause would involve the likelihood of arrest – the "hologram" was able to protect the individuals from such a fate!

At the other end of the size scale, we are seeing iPhone users encouraged to buy a new device which attaches to the screen in the form of an inverted pyramid. This appears to display an animated floating image when the screen itself displays an image from specialist software. Again, in this case, the reflection of the screen appears as a virtual image apparently within the volume of the pyramid, enabling the viewer to perceive a floating image from any oblique viewing position when the phone is laid upon a table.

1.11.5 Anaglyph Method

"Anaglyph" is a term which, like "hologram", is rooted in Greek origin, but which refers to relief carving. In modern terms, the use of viewing glasses with red and cyan lenses for the two eyes allows us to view printed images which have two stereographic views encoded in red and green ink. But the original print looks quite confusing when viewed without the glasses; the apparently "blurred" red, cyan and brown image is unsatisfactory to the unaided eye of the viewer.

The recent boom in 3D cinema projection has been achieved by a similar method. Here, the spectacles are a little more elaborate and comprise neutrally coloured polarising filters. The screen itself is made from a surface which retains the polarisation of the incident projected beam ("silver screen"). Circular polarisation instead of linear polarisation is used, as this is more sympathetic to members of the audience having the ability to tilt the head whilst viewing comfortably without disruption.

During my employment at 3M Research, Mike Fisher was a physicist in the photographic unit. He was one of the leading stereo photographers of that time and his twin-lens SLR camera had been custom-made from a pair of high-quality 35 mm film cameras. His accounts of sawing through two expensive SLRs to produce a single stereo camera enthralled our 3M coffee-room audience. He recorded colour slides in the camera and demonstrated a projection system with a twin-lens polarising projector involving remarkable lap-dissolve facilities. At that time, a reel-to-reel tape recorder provided a sound track and digital triggers to the projection system. Displays of the stereographic photography provided a remarkable experience for regular meetings of the company's technical forum. The London Stereographic Company was formed in 1854 and still exists under the directorship of Brian May and Elena Vidal [3].

Over the years there has been a range of commercial, purpose-made cameras for stereo photography, and although the Nimslo camera is no longer made, second-hand cameras with four lenses which would image sequences of 35 mm film are still relatively easily available.

In printed form, when a simple two-channel stereographic figure is presented in the form of twin images to the viewer's eyes in an appropriate manner, there is absolutely no difficulty in seeing a 3D image as a fusion of the pair of images without any viewing device.

Figure 1.14 shows such a pair of simple geometrical figures, which can be viewed autostereoscopically with a little effort. The viewer must position the stereo pair directly in front of the eyes and ensure that the left eye focuses solely upon the left print and that the right eye concentrates only upon the right print. This is done by staring into the distance. After a few seconds, the stereographic 3D image will appear to the viewer in

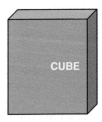

Figure 1.14 **Stereo.**

Figure 1.15 Barney – the one-eared Border Collie.

the void between the frames. Moving closer or farther from the page will enable the viewer to find an optimal position.

Now, the brain is telling the viewer that she sees *three* cubes, the central one of these is three-dimensional, and note that on *either side* of the 3D cube at the base is a logo "**Figure 1.14** Stereo". The logo "3D CUBE" that appears on the surface of the central perceived 3D image, shows the destructive effect of tilting the head whilst viewing the pair, as mentioned above with respect to 3D cinema.

Any suitable pair of photographs comprising sequential perspective will offer a three-dimensional view in this way, as shown in Figure 1.15. The position of the axis of rotation of parallax defines the extent to which the image appears to project forward.

The traditional anaglyph presentation is a graphic display in coloured inks which can be differentiated by coloured gels in viewing glasses. The binocular vision of the viewer is then able to ensure that each eye sees a slightly different picture without any need to train the viewer's eyes in a special viewing technique. The technique is easily applied to photographic reprint images, but, as mentioned, the problem for this technology is that the image tends to appear confusing without the viewing aid; bear in mind, however, that exponents of the technique could equally argue that holography appears confusing without suitable lighting!

These stereographic methods cannot realistically be called holograms – but, in fact, holography could help here in some cases.

A very simple two-channel hologram can easily be made where the viewer will effortlessly be able to see the 3D image in full colour provided that he or she keeps a central viewing position such that the left and right eyes are restricted by the hologram viewing properties to see only the appropriate information for the binocular stereo effect.

Of course, more complex holograms explained later can quite easily dispose of the need to restrict the viewing conditions as well as adding far more parallax data to the image. The limiting quality of the anaglyph methods is that only a single perspective is represented; unlike the holographic stereograms explained in Chapter 9, there is no variation in perspective of the 3D image as the viewer moves from side to side.

1.11.6 Lenticular Images

Plastic moulded lenticular screens comprise a thin layer of thermoplastic which is pressed with an array of adjacent cylindrical lenses. Various profiles for the lenses are possible. We see from the schematic in Figure 1.16 that each individual "lenticule" has

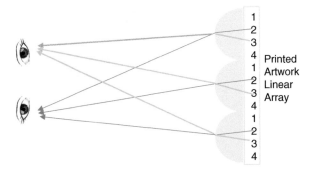

Figure 1.16 Mode of operation of lenticular devices.

the ability to guide a ray of light from any vertical line at its focal plane towards the field of view wherein a slitted window of view will be created.

Thus, the artwork is presented as a separate printed sheet, or could be printed directly on the reverse side of the thermoplastic array in the form of "slices" of view of a subject featuring different perspectives or animations. Each linear feature will create an adjacent window of view for the viewer's eye. The moment of alignment of a lenticular array sheet with corresponding printed or photographic artwork is an unforgettable experience, as the unconvincing, processed, flat artwork springs forth to be experienced for the first time as an integral 3D image. By the lenticular technique, we can organise for the viewer's eyes to experience two related views of the subject matter which may be arranged to convey stereographic or animation data to the viewer (or both). In Figure 1.16, the left eye sees the third channel of artwork across the scene; simultaneously, the right eye receives the image displayed in the second channel.

Using modern printing and lens embossing techniques, the designer can choose an appropriate quantity of channels in the printed matter to achieve the most effective result. A simple animation such as the security graphic in the British driving licence requires only two channels of information, whereas my co-author's portrait of David Bowie, shown in Figure 1.17, contains multiple channels recorded as a photographic sequence to enable the recreation of an animated 3D image.

1.11.7 Scrambled Indicia

Micro-lens arrays can also be produced in thermoplastics in much the same way as the lenticular method described above. Alfred V. Alasia, who died in 2010, founded Graphic Security Systems Corporation and used such lens arrays to produce the effect known as "scrambled indicia" [4] where a filter in the form of a lens array, when overlaid on a security document, was able to reveal to the viewer equipped with the correct viewing key, hidden information in a printed security device. This method was used in US postage stamps. The general technique also has the ability to produce 3D images. The lenticular and micro-lens array techniques are inferior to holography in terms of resolution capabilities, but, of course, are less demanding in terms of lighting conditions for reconstruction of an *apparently deep* three-dimensional image.

Figure 1.17 Animated lenticular portrait of David Bowie.

1.11.8 Hand-drawn "Holograms"

William J. Beatty published a document called *Hand-drawn Holograms* and this describes a method by which some intriguing 3D effects were created by a scribing technique.

This is based upon an effect with which we are all familiar. When driving the car with a windscreen which refuses to clear adequately, the motion of the wiper blades tends to smear dirt in curved lines across the screen. Similar effects appear on the car bodywork if a polishing cloth contaminated with grit is used so as to scratch the paint surface as the cloth sweeps across the paint in a curved line.

If these patterns are viewed by the light of street lamps, we perceive three-dimensional curved lines appearing to sweep into the distance and a repetition of their rotational motion as we drive past successive light sources.

Beatty harnessed this effect in such a way as to encode the scribed pattern by using a pair of dividers with a sharp stylus point to gently score the surface of a shiny plastic or metal sheet with a sequence of circumferential lines centred on the locus of a simple image such as an alphanumeric character. This method is described [5] at http://amasci. com/amateur/holo1.html

I achieved excellent results by following this description and the result is a white, three-dimensional image. The key is the action of the stylus of the dividers utilised on

(a)   (b)

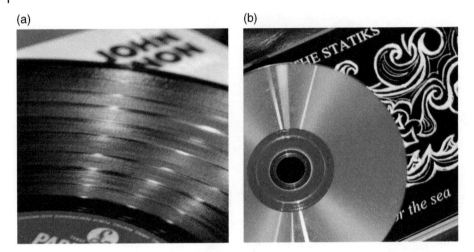

Figure 1.18 (a) Vinyl record grooves; (b) CD pit tracks.

the plastic layer. Gentle manipulation of the stylus point tends to burnish a smooth curved v-groove with an angled reflective edge in the plastic layer.

During viewing, the point source of light is reflected from a single point of one of the curved v-shaped groove edges towards the viewer. If the viewer moves, then this highlighted position moves accordingly. A v-shaped curved groove representing an arc centred on successive adjacent points tracing the graphic which is the subject of the intended image will, of course, present a slightly different highlight, and the integral effect of all these curves is to represent a function of the original subject; with the ability to include three-dimensional effects.

The white reflected image in this method results from the fact that this is a non-diffractive technology. It is an interesting opportunity to compare in this way the effects we see from the reflection from the surface relief associated with a vinyl disc and a CD recording, as shown in Figure 1.18(a) and (b), where the line spacing gradually approaches the frequency that we associate with holograms and where diffraction of the visible wavelengths begins. There is a hint of colour in the light reflected from the vinyl grooves in (a) and in (b) a distinct rainbow effect with which we are all familiar. This is because the spiral track of pits on a CD has spacing of $1.6\,\mu$, whereas the vinyl record grooves are of the order of 100 times greater spacing.

1.11.9 "Magic Eye"

In the 1990s a company was formed under the banner N.E.Thing Enterprises Ltd to exploit the technique of single image random dot stereograms (SIRDS) which is credited to Hungarian expatriate Béla Julesz working at Bell Laboratories in the 1950s (simultaneously with laser pioneers Townes and Schawlow).

During that era, the 3D images achieved cult interest in the Far East and millions of books were sold. The company later changed its name to Magic Eye Inc.

The books, such as *Magic Eye* [6] provide a non-technical, illustrative range of images using the SIRDS technique.

As we saw with the "stereo pairs" in Section 1.11.5, the method demands that the viewer controls their eyes in such a way as to stare "through" the printed image. This effectively creates two independent channels of image data, which differ slightly. The images transmitted to the brain of the viewer from the left and right eyes contain lateral displacement for certain groups of dots or vector components. As a result, the brain interprets these differences as depth cues.

The result is that it is relatively easy to produce 3D scenes that typically comprise a patterned rear plane which forms a background to a planar or three-dimensional feature in the mid-ground, behind the surface of the printed paper.

Notes

1 Rothman, T. (2003) *Everything's Relative and Other Fables from Science and Technology.* John Wiley & Sons Inc., Hoboken, New Jersey (ISBN: 978-0-471-20257-8).
2 www.smitherspira.com "Ten-Year Forecasts of Disruptive Technologies in Security Printing to 2020."
3 The London Stereographic Company. www.londonstereo.com
4 Graphic Security Systems Corporation. www.graphicsecurity.com
5 William J. Beatty http://amasci.com/amateur/holo1.html
6 N.E. Thing Enterprises (1993) *Magic Eye: A New Way of Looking at the World* (ISBN 0-7181-3804-X).

2

Important Optical Principles and their Occurrence in Nature

With the cooperation of Dr Alexandre Cabral

Laboratory of Optics, Lasers and Systems and Department of Physics, Faculty of Sciences, University of Lisbon

2.1   Background

During many years of work in commercial, artistic and academic holography, the authors have become aware of a significant technical information barrier to interest and involvement in making a hologram, for a wide range of people who have a potential interest in holography. This category includes not only students in scientific disciplines, but students of art, design and graphics as well as laboratory technicians, security specialists and, indeed, sales and marketing executives who wish to have a practical level of understanding of how some of the iconic holograms of our time have been made. This barrier to entry is exacerbated by the frustration of opening many a textbook to reveal pages of advanced mathematics, physics, chemistry and graphics principles which are beyond automatic comprehension for many potential enthusiasts.

It is our intention in this book to take into account only the immediate scientific principles that impact directly upon our ability to understand the workings of the holographic studio and the components and principles which enable us to make successful holograms. Readers who find areas of special personal interest here may wish to get further detailed background information on the behaviour of light from the fifth edition of Eugene Hecht's *Optics* [1].

By offering our abridged explanation of the physics and optics required for holography, we hope to assist in the widespread growth of a technology that has brought us so much pleasure, interest and the opportunity to take part in the birth of a fascinating industry, which has expanded on a worldwide scale with every advance in light science. At this moment we face new developments in film and laser technology which will provide better access to holographic technology for a far wider range of people.

We all experience the effects in our everyday life of the optical principles which we are about to discuss. There are events we shall consider, such as opening our eyes whilst swimming under water, which defy our ordinary visual experiences, but also really common experiences such as the elusive varying colours of a butterfly as it flutters over the garden pond, that have attracted attention since time immemorial, but only in recent years has the mechanism for this display been fully understood.

The Hologram: Principles and Techniques, First Edition. Martin J. Richardson and John D. Wiltshire.
© 2018 John Wiley and Sons Ltd. Published 2018 by John Wiley & Sons Ltd.
Companion website: www.wiley.com/go/richardson/holograms

Figure 2.1 Blue Morpho butterfly.

1 cm

In 1969, in the Strawbs song *Or Am I Dreaming?* [2] Dave Cousins observed:

> The fragile gentle butterfly
> With multi-coloured wings
> Settles on the toadstools
> In the midst of fairy rings

At that time, hearing these words, I might have presumed that pigment was responsible for the beautiful colour of a butterfly; but with advances in modern microscopy, including laser illumination, it is now firmly understood that pigment is absent in many such instances, and it then becomes very clear how a *diffractive* mechanism gives rise to subtle changes in the colour as lighting conditions and viewing angles change. A structure of minute cells means that the diffractive microstructure of the wing surface is responsible for the intense colouration of the Blue Morpho butterfly in Figure 2.1.

Advanced academic studies have been carried out into the microstructures which yield these optical effects [3].

As soon as we begin to study the detailed physics behind such natural phenomena, we can recognise the course by which the early experiments in physics and optics developed in the perceptive minds of incredibly observant pioneers like Newton, Huygens, Galileo, Young, Brewster, etc., from everyday visual experience using natural sunlight; and how we arrived at the level of understanding of the physics over hundreds of years which has resulted in our ability today, to make holograms and other optical devices reliably, as a matter of course.

In fact, the garden pond which is the environment of the previously mentioned butterfly, provides superb illustrations of many other optical effects, which, when examined more closely, start to explain many aspects of the operation of the holography laboratory.

- A child kneeling beside the pond leans over to peer into the water. She is amused to see her own reflection and notices that the reflected face appears as far below the water as she is above the surface. (*Virtual image and the laws of reflection.*)
- The rigid branch of a plant beside the pond extending into the water, which appears to bend sharply at the interface of air and water. (*Snell's Law.*)
- The "rainbow" colours seen as the sun shines on the water droplets from a fountain re-circulating water in the pond. (*Dispersion.*)
- Another plant whose movement in the wind causes a branch to touch the surface of the water gently, with the effect that circular waves are emitted from the point of contact. (*Wave phenomena.*)

- The continuous rise and fall of a leaf in the midst of that activity shows the general ability of any form of detector to record a wave event. In this natural environment, the property of a leaf; but in our laboratory, the property of silver halide photo emulsion, photopolymer or photo-refractive crystal. (*Phase recording.*)
- The introduction of another nearby branch will create another wave centre and at the point of intersection of these waves, we will see a much more complicated pattern of interference between the two wave centres. This really is the doorway to our ability to create holograms from waves, in the form of light waves – the electromagnetic radiation which we are able to harness with such expertise in the modern era. (*Interference.*)
- It does not take much imagination to envisage that if we were to replace the aforementioned floating leaf bobbing upon the waves in the pond with a large number of artificial floats attached to a mechanical system capable of replicating their movement, then the mechanical repetition of their movement sequence is highly likely to recreate some form of "image" of the original disturbance. (*Image reconstruction.*)
- Sunlight reflecting from the water surface is predominantly plane polarised. Thus, our ability to use "Polaroid" glasses to enhance our view of the fish in the pond is well known, because a polarising filter (analyser) can selectively eliminate much of the light reflected from the water surface and thus allow an improved perception of light coming from sources within the pond. Plane polarisation is a vital property of the laser light we use to record a hologram. (*Brewster's Law.*)

2.2   The Wave/Particle Duality of Light

Although Aristotle implied that light was "a disturbance of the aether", it was first seriously proposed in the seventeenth century by Christiaan Huygens, contemporary with Newton's preferred particle theory, that light (electromagnetic radiation) propagates in the form of a transverse wave – a wave whose electrical and magnetic field oscillations are perpendicular to the direction of propagation (Figure 2.2).

Figure 2.2 Magnetic and electric fields in transverse oscillation. Figure courtesy of Alex Cabral.

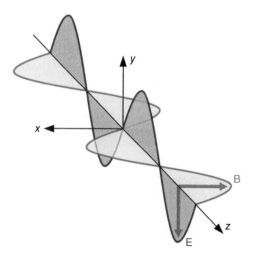

Whereas we rely upon the concept of the photon as a discrete package of light energy, we acknowledge simultaneously the Huygens wave concept and this has stood the test of time admirably, including its compatibility with modern quantum electrodynamics.

In fact, holography is a classic example of the usefulness of the wave model, since, in essence, we are using its technique to "re-build" original wave patterns. Simultaneously, our explanation of the operation of the lasers which facilitate our work is simplified significantly by the assumption of the existence of the particulate photon.

The acceptance of twin explanations of the behaviour of electromagnetic radiation is known as *wave/particle duality*. Whereas the discussion of the precise nature of light has endured since the days of Newton and Huygens, it is, of course, interesting to reflect that in much more recent times, a similar duality has been firmly established in the description of electrons and other phenomena previously regarded categorically as particles. The situation nevertheless remains complicated and a matter for debate amongst leading physicists, and Einstein himself doubted whether our explanation is yet complete.

Basically, we need to be able to accept that the combination of two concepts of existence for electromagnetic phenomena allows us to explain all of the effects that we experience in nature and in the optical laboratory, and we apply the most appropriate explanation to each individual situation.

Since the original Huygens analysis did not necessarily cope with general diffractive effects independent of wavelength, the contribution of Augustin Jean Fresnel at the turn of the eighteenth century was to modify the principle with qualifications which allow for diffractive effects.

According to wave theory, electromagnetic radiation manifests itself in the form of a sine or cosine wave, so the relationship with a wave on the water surface is clear. In Figure 2.2, the electric vector is the upright wave and its sinusoid profile is in exact phase with the perpendicular magnetic vector; both of the vectors are transverse to the direction of propagation. The plane of polarisation is regarded as the plane of the electric vector. Linear polarisation is the simplest form for an electromagnetic wave. A more general condition is elliptical (and circular) polarisation, which can occur in the presence of birefringent materials or certain other unusual optical configurations.

2.3 Wavelength

An electromagnetic wave is periodic in both space and time. In Figure 2.3, the spatial period of the wave between adjacent maxima is the wavelength, λ, of the radiation. A similar description can be made for the temporal period, T, and both periods are related with the expression: $\lambda/T = c$, where c is the speed of light.

The same expression can therefore be written considering the frequency of the radiation ($\nu = 1/T$) in the form:

$$\lambda \times \nu = c$$

The wavelength range of electromagnetic radiation is a continuum, of which the range of visible wavelengths which we utilise for holography (as a result of our visual perception) is a *very* small central portion of the whole spectrum, as illustrated schematically in Figure 2.4, which shows the approximate areas occupied by many of the

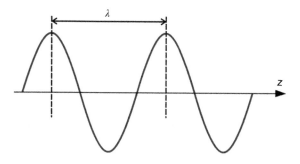

Figure 2.3 Electromagnetic wave period. Figure courtesy of Alex Cabral.

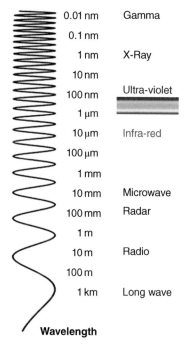

0.01 nm	Gamma
0.1 nm	
1 nm	X-Ray
10 nm	
100 nm	Ultra-violet
1 μm	
10 μm	Infra-red
100 μm	
1 mm	
10 mm	Microwave
100 mm	Radar
1 m	
10 m	Radio
100 m	
1 km	Long wave

Wavelength

Figure 2.4 Electromagnetic spectrum. Figure courtesy of Alex Cabral.

forms of electromagnetic energy with which we are relatively familiar in everyday life, without necessarily appreciating the vast range of orders of magnitude in their wavelengths.

But it is logical that many of the rules governing the properties of light that we manipulate for our needs in holography also apply in parallel technologies at other wavelengths.

For example, my own first detailed practical experience of the technicalities of diffraction, which clearly is the basis of holography and a vital topic in this book, came in my work at 3M Research in the use of x-ray diffraction techniques in the exploration of silver halide crystals, utilising the principles of Bragg diffraction, which we seek to harness to our advantage in the everyday process of hologram production.

There is a general, advantageous trend in the spectrum which we should bear in mind. The shorter wavelengths of electromagnetic radiation are associated with more energy per photon than the longer wavelengths. In everyday life, we therefore tend to be far more apprehensive about the hazards of x-rays than the ubiquitous radio waves around us, and without concern, we have powerful microwave emitters in our kitchens and our pockets!

Planck's equation accounts for this inequality:

$$Q = h\nu = hc/\lambda$$

Where Q is photon energy (joules), h is Planck's constant, c is the speed of light, λ is the wavelength and ν is the frequency.

This linear relationship is very useful in our attempts to measure light flux density on the holographic table; we are often measuring very small levels of energy and the single detector head of the light meter is able to cope with the whole range of the visible spectrum, principally because we are able to calibrate its thermal detection of the laser light simply by introducing wavelength information prior to taking a measurement.

The laser is a very special case in the discussion of the concept of wavelength. In this case, we see three properties which differentiate laser light from other light sources:

1) A highly *monochromatic* light source;
2) A spatially and temporally *coherent* beam of light;
3) An unusually *highly collimated* beam.

When a laser cavity is used to amplify light of a single wavelength to high levels of output, we experience an unusual sight which is not duplicated elsewhere. When the laser cavity selection process for the bandwidth of the light results in such a narrow peak of emission, then it is possible to achieve a situation where every wave is in phase and we refer to finite "coherence length" and a visual appearance of absolutely "pure" colour which we rarely experience in everyday life (with the exception of arc lamps, such as the sodium arc lamp used commonly in street lighting).

In the laser laboratory, it is possible, with experience, to be able to differentiate quite easily by eye single-line wavelengths of close proximity such as common reds (647 nm and 633 nm) or blues (488 nm and 476 nm).

2.4   Representation of the Behaviour of Light

2.4.1   A Ray of Light

In the study of optics, it is necessary to represent the passage of light from one point to another in the form of a ray of light. But this concept is simply a tool in our method of study. In practice, the ray of light is not a physical entity. If we consider a laser beam as a first example, it is possible to imagine a very narrow beam, and the concept of a ray of light could be considered to be the infinitesimally thin limit to such an assembly of waves or photons.

2.4.2   A Wave Front

This is the surface which represents the locus of positions of identical phase in an assembly of light waves. When we begin to look at the various laws of optics, we are considering the effects of various phenomena upon the wave front, and in this case we need to consider the properties of the "wave front" rather than the "ray of light" to account for their properties. In an isotropic medium, however, there is no temporal change to the linear progress of the wave front, and this effectively gives rise to the value of the concept of the light ray as the straight line connecting points of optical activity. The simplest example of this concept is the spherical wave front existing after the focus from a lens, with its radius equal to the distance from the focus.

2.5   The Laws of Reflection

Reflection is, in many ways, the easiest optical phenomenon to understand. This applies to the incidence of light upon a smooth surface, such as a polished metal, as shown in Figure 2.5.

Figure 2.5 Reflection at a flat surface. Figure courtesy of Alex Cabral.

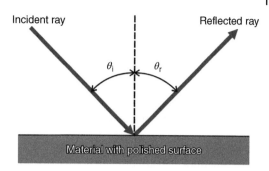

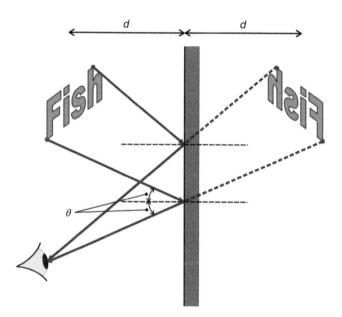

Figure 2.6 Reflection in a plane mirror. Figure courtesy of Alex Cabral.

As a first example, it is simple to consider perfectly flat surfaces only. There are two simple laws:

1) The incident ray and reflected ray are in the same plane normal to the surface.
2) The angle of incidence, θ_i, is equal to the angle of reflection, θ_r (in optics the angles are measured relative to the normal; this tends to differ from engineering practice).

The reflection in a plane mirror (or the surface of a pond) produces a virtual image for the observer. This image appears to the viewer to be positioned an equal distance, d, behind the mirror surface as the object is in front, shown in Figure 2.6.

In the case where the surface is curved, we may still imagine the surface as a continuum of tangential flat surfaces, with the result that the laws of reflection remain valid. In this case, the phenomenon of magnification is introduced into the virtual image.

2.6   Refraction

The Oxford English Dictionary defines refraction as:

> The fact or phenomenon of light, radio waves, etc. being deflected in passing obliquely through the interface between one medium and another or through media of varying density.

Essentially, this is a function of the difference in speed at which light travels in materials of varying optical density.

2.7   Refractive Index

The oft-quoted "speed of light", c, of approximately 3×10^8 m/s refers to its absolute speed in vacuum, which clearly has an immensely important significance in physics. In other media, however, the progress of the electromagnetic wave is impeded by its interaction with the atoms and therefore light travels more slowly.

Of course, we can say that the velocity in a low-density medium, such as air, is reduced very little, so that, for holography purposes, we can assume the refractive index, *n*, of air to be 1.000 (at standard temperature and pressure [S.T.P.] it is, in fact, 1.00029).

Traditionally, refractive index is measured for the sodium D-line, which is the pair of orange/yellow emission lines at 589 nm as shown in Chapter 3, with which we are all familiar in the sodium vapour lamp.

So this number, *n*, is the ratio of the speed of light, *s*, in any transparent material divided into the absolute speed of light, c, in vacuum:

$$n = c/s$$

As a ratio, it has no units.

It is very important to realise that, although this parameter is a ratio assessed at a single wavelength as described above, its absolute value varies to some extent with the wavelength of radiation, which, as we shall see, is the reason for familiar dispersive effects.

2.7.1   Refractive Index of Relevant Materials

In practice, the calculations we need to make routinely for holography purposes require us to think only in terms of the refractive index of a limited list of useful materials in the laboratory (see Table 2.1).

2.8   Huygens' Principle

This is the original observation in 1678 that "any point on a wave front of light may be regarded as the source of secondary waves and that the surface that is tangential to the secondary waves can be used to determine the future position of the wave front."

This assists us in visualising the passage of light through the boundaries of dielectric media. The Huygens principle followed on from Fermat's 1662 statement that light will

Table 2.1 Refractive index of common materials.

Air	1.00
Crown glass	1.50
Water	1.33
PET film	1.65
Triacetate film	1.48
Perspex acrylic	1.50
Methanol	1.33
Decalin	1.48
Ethanol	1.36
Cedar oil	1.49
White spirit	1.43
Iso-propanol (IPA)	1.38
AgHa/Gelatin emulsion	~1.64
Silver bromide crystal	2.25
(c.f. Diamond)	(2.42)

follow a path between two points which takes the minimum period of time. This is a general principle of nature which is associated with the conservation of energy.

2.9   The Huygens–Fresnel Principle

With the qualification from Fresnel in 1816, the principle also helps to explain diffraction at an aperture: "every *unobstructed* point in a wave front, at a given instant in time, serves as a source of secondary spherical wavelets of the same frequency. The amplitude of the optical field at all points forward is the superposition of all these wavelets including their phase and amplitude".

As defined above, a wave front is a surface *in which light waves have a constant phase.* Thus, the atoms in a dielectric medium through which light is transmitted are regarded as producing secondary wavelets and the nature of those atoms will dictate the rate at which the radiation proceeds. Of course, in vacuum there are no atoms to influence the wave front, and no diminution of speed.

Without delving any deeper than we have into complex physical and mathematical aspects of wave theory, it is sufficient to acknowledge that the varying speed of light in different media means that, at an interface between air and a more dense medium such as a glass plate, an incident wave front arriving obliquely at the surface will be redirected simply because, as shown in Figure 2.7, part of the wave front will spend more time travelling in the less dense medium (air), and will thus progress more quickly. The flank of the wave front which arrives at the interface first progresses less quickly thereafter, until such time that it arrives at the second boundary, where its path will redirect again.

Huygens' principle dictates that "the surface that is tangential to the secondary waves can be used to determine the future position of the wave front." The effect is that the locus of these tangential representations of the phase of the array of waves

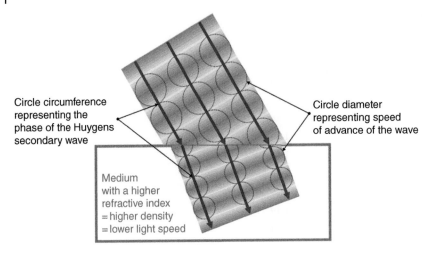

Circle circumference representing the phase of the Huygens secondary wave

Circle diameter representing speed of advance of the wave

Medium with a higher refractive index = higher density = lower light speed

Figure 2.7 Behaviour of the wave front at a refractive index boundary. Figure courtesy of Alex Cabral.

demonstrates the effective change of direction of the wave front, rotating towards the normal as the light enters the medium of higher density.

For a parallel-sided glass plate (such as a hologram recording plate), the action at the second surface will represent an equal and opposite redirection, so that the emerging ray will be parallel with the incident ray whilst slightly displaced laterally. In the case of a prism, the action at the two surfaces is *not* equivalent and the transmitted ray will travel in a modified direction; different wavelengths of light are thus dispersed.

The general concept of the effect of the speed of movement changing with the containing medium is illustrated in everyday terms by imagining the example of a person challenged to rescue a swimming dog in distress (Figure 2.8). Fermat's principle, expressed for the passage of electromagnetic radiation, coincides with the need of the rescuer on a beach to reach the drowning dog in the shortest possible time. The lady can run faster on the beach than she can swim in the water – there is therefore only one optimal route to achieve this. A person who can run at the same speed, but swim less quickly, should follow a different path to reach the animal in minimal time. But a person who is able to swim at exactly the same speed that they run should follow a straight line – this equates to the effect in optics where there is no change in refractive index in a beam path.

2.10   Snell's Law

In 1621, Willibrord Snell perfected Kepler's original explanation of the behaviour of light arriving at the interface of two differing media. His contemporary, Descartes, was the first to publish and the work is sometimes called the Snell–Descartes Law.

Each time we shoot a hologram, the laser beam entering the glass or film substrate is deflected at the interface toward the normal, as shown in Figure 2.9, in a way predicted by Snell's Law in accordance with the equation:

$$n_i \sin\theta_i = n_t \sin\theta_t$$

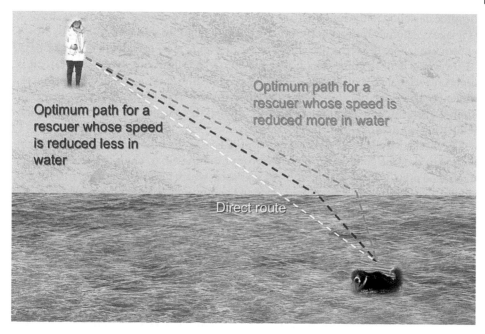

Optimum path for a rescuer whose speed is reduced more in water

Optimum path for a rescuer whose speed is reduced less in water

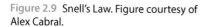

Direct route

Figure 2.8 Practical example of the effect of travelling in different media.

Figure 2.9 Snell's Law. Figure courtesy of Alex Cabral.

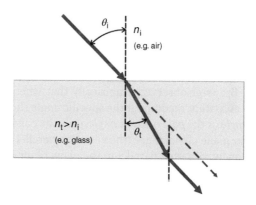

where n_i is the refractive index of the first medium from which the ray is incident;
θ_i is the angle at which the ray arrives;
n_t is the refractive index of the second medium;
θ_t is the angle at which the transmitted ray proceeds in the second medium.

In optics, angles are conventionally positive when measured anticlockwise relative to the normal.

In practice, in the holography lab, the glass or film substrate is present upon both recording and reconstruction of our recording, so that we do not necessarily, in general, need to take account of its effect.

In fact, when the recording plate is put into place, it does modify the path of the beams. But as soon as the processed holograph is returned to its position in the reference beam, the same deviation of the beam path re-occurs and in the presence of the plate substrate, its image is naturally formed in the precise position that it was recorded; there is, thus, a tendency to ignore the refractive effects of the recording plate or film in the general case.

However, when we come to consider the *precise* details of our fringe structure, it is important to realise that the effects of the glass or film may, in some circumstances, influence the fringe microstructure recorded. In this way, if we allow shrinkage, or attempt to shrink or expand the *layer*, it is prudent to consider the precise orientation of the fringe structure with respect to that layer.

We can see that the most typical situation we face in the hologram mastering system is the arrival of the collimated reference beam from the air, to enter the glass plate. As mentioned previously, for holography we use lasers which provide linear polarised light, and the plane of linear polarisation is arranged to prevent unwanted internal reflections in the glass substrate. The influence of polarisation upon surface reflections was originally discovered by Sir David Brewster, Professor of Physics at St Andrews in the nineteenth century.

2.11   Brewster's Law

Fresnel's equation (see Section 4.3.2 in the second edition of Hecht's *Optics*) established that when a beam of incident light is polarised in the plane of incidence upon a glass surface, at the angle where no light is reflected, *then the angle between the reflected and transmitted beams will be 90°.*

The E-field of light in the glass under these conditions is thus parallel with the direction of the prospective reflected wave and, therefore, cannot propagate in that direction. (Hecht Section 8.6 shows the implications of this fact in terms of electron-field oscillation).

Brewster observed empirically that when an un-polarised wave is incident upon a transparent medium at the specific angle shown in Figure 2.10, now called the *Brewster angle* or the *polarisation angle*, only the component which is polarised perpendicular to the plane of incidence (the vibration parallel with the surface) will be reflected; the component vibrating parallel to the plane of incidence will be transmitted into the glass.

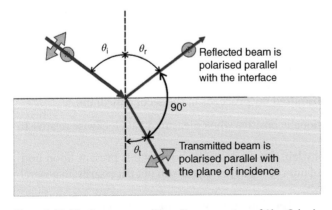

Figure 2.10 The Brewster condition. Figure courtesy of Alex Cabral.

This explains the success of "Polaroid" sunglasses under certain circumstances; these selectively eliminate from view the light reflected from a water surface and thus permit us a better view of the generally un-polarised light actually emerging from a pond through the interface of air and water.

This principle is particularly useful to us on the optical table because when we are using linear polarised light, which almost by definition is required for hologram production, we can achieve 100% reflection or transmittance, and thus, in some cases, ensure that we avoid undesirable surface reflection and corresponding "wood grain" defects (representing thickness variations of the glass) in our recordings.

Furthermore, we shall see later in this book that the "opposite" effect is utilised inside the laser cavity itself, in the form of a *Brewster window*, which selectively reflects cross-polarised light *out* of the cavity and is therefore able to ensure the vital production of a linear-polarised laser beam. Importantly, the phenomenon of interference, at which we shall look in detail later, is dependent upon the light waves involved having a similar plane of polarisation.

From the diagrams, the Brewster angle for a glass/air interface can be calculated as follows:

If glass has refractive index $n = 1.50$ and air $n = 1.00$

$$\text{Brewster angle is } \theta_p = \arctan\left(n_2/n_1\right)$$

$$\theta_p = 56.3°$$

From Snell's Law:

$$n_i \sin\theta_i = n_t \sin\theta_t$$

$$1.00 * \sin 56.3° = 1.50 * \sin\theta_t$$

$$\theta_t = 33.7°$$

In terms of simple trigonometric calculation from the diagrams, since the angle of incidence is equal to the angle of reflection, and the angle between the refracted rays is a right angle, in accordance with the Fresnel condition, the angle of the transmitted ray, θ_t, is confirmed as:

$$\theta_t = 180° - 56.3° - 90°$$

$$\theta_t = 33.7°$$

In the specific case of hologram production, it immediately occurs to us that a ray of incident light therefore proceeds into the recording plate at a reduced angle (rotating towards the normal) at the interface, as shown in Figure 2.11; and *within* that more dense medium, interference with another incoming "object beam" will, in fact, occur at that particular modified axial angle. However, we will explain in Section 2.18 why there is no reason in our holographic application to be concerned by such effects, and justify our freedom to overlook them in our everyday design calculations.

Similarly, in accordance with Table 2.1, the refractive index, n, of a photo emulsion is reasonably close to the index of the glass plate itself, so it is also reasonable, for simplicity, to overlook any refraction or reflection which occurs at the gelatin/glass interface when we are considering typical hologram production.

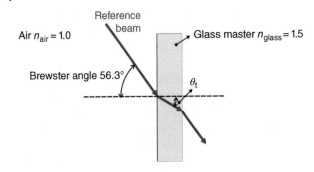

Figure 2.11 Brewster angle incidence at a glass plate. Figure courtesy of Alex Cabral.

2.12   The Critical Angle

From Snell's Law there is an obvious limiting case which is of interest. This is where Snell's Law predicts a situation where light is unable to exit the more dense medium; when the angle of incidence exceeds the so-called *critical angle*.

We rarely notice this effect, because with the exception of our experience when swimming under water, our eyes are routinely on the low-density side of an optical interface; there is no critical angle effect where light is incident on an interface from the low-density medium.

However, when submerged, we have an example of the manifestation of the "critical angle" effect; light arriving at the water/air interface from the bottom of a swimming pool is able to exit the water at only a limited range of angles. We notice this limitation when submerged if we look up from the bottom of the pool, because we appear to see through the interface only in a limited area of the surface of the water above us.

From more oblique angles, the internal reflection of light at the surface then tends to exceed the intensity of light crossing the interface. In Figure 2.12, the camera appears to capture only a limited area of the dark exterior above the pool. However, in reality, we see from the ray diagram in Figure 2.13 that rays of light from objects over a wide viewing angle are introduced into the water from the less-dense medium. This is sometimes called *Snell's Window* and is actually the explanation for the common term *fish-eye lens* in photography. The remainder of the scene in the photograph in Figure 2.12 [4] comprises the brightly lit internal reflection of the pool base in the water/air interface as a direct reflection from the relatively still water surface. This phenomenon is called *total internal reflection (TIR)*.

In this case,

$$n_i \sin\theta_i = n_t \sin\theta_t$$
$$1.33 \sin\theta_i = \sin 90$$
$$\sin\theta_i = 1.00/1.33$$

The critical angle for an interface between water and air,

$$\theta_i = 48.6°$$

Figure 2.12 Underwater view from a swimming pool [4].

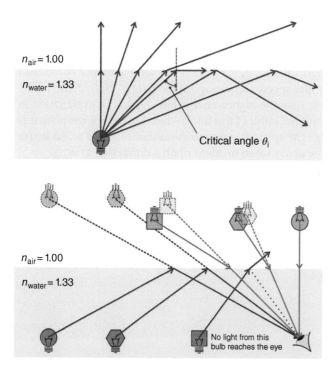

Figure 2.13 Ray diagrams explaining underwater view in a swimming pool. Figure courtesy of Alex Cabral.

2.13 TIR in Optical Fibres

A vital, modern manifestation of total internal reflection arising from the critical angle principle is the optical fibre. Light projected into a thin glass filament is unable to escape at the boundary of the glass fibre.

In the doped (graduated index) fibres I made during my employment at Standard Telecommunication Laboratories in the early years of optical fibre research, I found one of the most fascinating features of the fibre extrusion process to be the fact that the light from the furnace used to melt the graded index "pre-form" rod was used to monitor the efficiency of the pulled fibre. On the extrusion machine, the leading end of the fibre was inserted into a photodetector on the extrusion winding drum and was used to monitor the light from the furnace in real time as the fibre was pulled to a length in excess of 1 km.

Any fault (such as a bubble) in the fibre was then responsible for a sudden drop in the transmission of light to the detector. Graded index fibres without any such fault were capable of transmission levels better than 99% over 1 km.

2.14 Dispersion

Since refractive index varies very slightly with wavelength, red light will deviate less than high-energy blue or violet light when Snell's Law is applied to behaviour at an interface when oblique incidence occurs. When we deal with parallel-sided glass optics, such dispersion is effectively reversed at the second surface, but of course in asymmetric optical components, such as a prism or a simple lens, dispersion of white light into its components will occur, and laser beams of various wavelengths will perform slightly differently.

Whereas light of all wavelengths travels in vacuum at the same absolute speed, and thus effectively in air at very similar speed, the passage of red light in more dense media (water, glass, etc.) is delayed less than the higher-energy blue light. The explanation in quantum mechanics is far beyond the remit of this book – but, for the holographer, it is more important to acknowledge the apparent contradiction that red light is *less active* at a refractive interface but *more active* when incident upon a diffraction grating!

If the refractive index of crown glass varies between red and blue light by approximately just $\Delta n = 0.02$, then we will see from the extension of Snell's Law above that the angle of dispersion of white light is of the order of only half a degree when a white beam is incident on the first surface of a prism at 45°, as in Figure 2.14(a).

For red light:

$$n_i \sin\theta_i = n_t \sin\theta_t$$
$$1 \times \sin 45° = 1.51 \times \sin\theta_t$$
$$\sin 45°/1.51 = \sin\theta_t$$
$$\theta_t = 27.9°$$

For blue light:

$$n_i \sin\theta_i = n_t \sin\theta_{t+}$$
$$1 \times \sin 45° = 1.53 \times \sin\theta_t$$
$$\theta_t = 27.5°$$

(a)   (b)

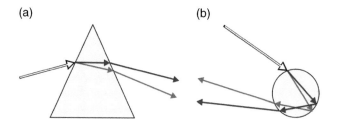

Figure 2.14 Dispersion of white light by (a) a prism and (b) a raindrop. Figure courtesy of Alex Cabral.

When this dispersed light arrives at the second interface of the prism, the colour components continue to disperse further. This differs from the case with which we are familiar on an everyday basis when light passes through a window. In that case, only a very slight separation is detected in the form of slight colour fringing at the upper and lower edges of a white beam, because the light is refracted in the opposite direction as it passes once more into the lower-density medium at the second surface of the glass.

Because different types of glass vary in refractive index, this minor difference in behaviour between various wavelengths allows lens manufacturers the opportunity to create "achromatic" doublet lenses.

In the case of a spherical raindrop, whose refractive index is about 1.33 in sodium light, we see in Figure 2.14(b) the explanation of the appearance of a rainbow. Light is dispersed in accordance with the same principle of increased index for the blue component, and, after internally reflecting at the back of the droplet, is refracted again at the surface. The viewer perceives the rays of light arriving at their eye and naturally envisages the "rainbow" to be displaced from the actual source of dispersion. In practice, the angle between the incident sunlight and the refracted light is typically of the order of 40°.

Colour dispersion is a significant problem in lens manufacture. In an achromatic doublet lens, the required magnification is achieved by the use of two component glasses of differing refractive index. At their common surface, the positive and negative lenses are cemented together. The effect is that lens manufacturers can achieve the required focal length with a much reduced level of chromatic aberration. Further explanation can be found in the first edition of Hecht (Section 6.3.2).

2.15   Diffraction and Interference

When two or more waves of coherent monochromatic light are travelling together in phase, their amplitude will add together where they overlap in space, as shown in Figure 2.15 (where, for simplicity, waves have the same amplitude).

But if they are in phase opposition, there will be destructive interference. Any condition in between these will give rise to a fading amplitude wave. The flux density of the light is equivalent to the square of its amplitude, so we see that we have the capability to distinguish easily in our recording material the zones (fringes) of constructive and destructive interference.

Light can be regarded as a scalar for our purposes, so that incident monochromatic coherent wave fronts of electromagnetic radiation arriving at our recording zone in

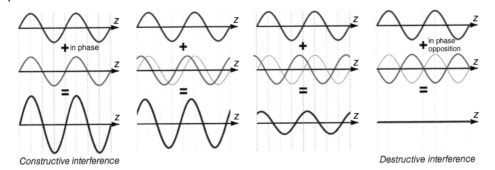

Constructive interference   Destructive interference

Figure 2.15 Diminution of constructive interference with phase change. Figure courtesy of Alex Cabral.

Figure 2.16 Diffraction at a small aperture.

conditions of physical stability with appropriate laser coherence will provide an eminently recordable standing wave for a hologram microstructure.

The term "diffraction" was coined by Francesco Grimaldi in the seventeenth century. It refers generically to the range of phenomena which occur when light encounters an obstacle, or an aperture of similar dimensions to its wavelength.

We have previously discussed the experiment by Thomas Young which was used to support the wave theory of light. Two narrow slits in a screen were able to allow light through in such a way that the light from each of the slits was able to interfere with light from the other.

In the simplest case, if we consider laser light from a single pinhole in a barrier screen falling upon a viewing screen farther upstream from the small aperture, as shown in Figure 2.16, there is a distinctive pattern of concentric rings projected upon the screen. We are quite familiar with this sight as it tends to occur every time we set up a spatial filter in the holography laboratory, at least until we achieve the objective of focusing the beam precisely into the pinhole aperture, at which point much of the edge diffraction is eliminated.

It is possible, by examining the pattern on such a distant screen, to predict the details of the aperture, for example, whether we use multiple small pinholes, a narrow slit or multiple slits, or even a photographic transparency.

In some cases, eliminating such edge effects is a significant problem in the making of a hologram; for example, masking down spread laser beams on the optical table will frequently demand considerable time spent in eliminating problematic edge effects in the recording itself, whilst failure to mask the table thoroughly will typically result in deleterious spurious images being recorded in addition to the required subject, with frustrating efficiency!

2.16   Diffraction Gratings

A diffraction grating is an assembly of very fine lines and, in accordance with the above general definition of diffraction, it is necessary to use very high resolution mechanical or holographic processes in order that the spacing of the individual line pairs is of the order of the wavelength of light.

Long before it was possible to make diffraction gratings by holographic means, it was recognised that it was possible to scribe "ruled gratings" mechanically by precision ruling engines. Rittenhouse, in the eighteenth century, and later Fraunhofer had independently produced crude gratings by using fine wire threads suspended in arrays.

In recent years, we have all become familiar with the "rainbow" of colour reflected from the surface of CDs and DVDs. In this case, we are seeing the incidental production of an efficient diffraction grating where the surface-relief microstructure resulting from the pits which represent the digital recording signal, happen to be suitably spaced in order to diffract and disperse incident white light efficiently into its component colours. The spacing of the individual surface features, the *period of the grating*, is related to the angle of diffraction and the wavelength of the light by *the grating equation*.

2.17   The Grating Equation

All diffraction gratings rely upon interference between wave fronts, caused by the microscopic diffraction lines whose dimensions are of the order of the wavelength of the electromagnetic radiation in question. The grating can be of the reflecting type or the transmitting type (for simplicity, we will describe the reflecting case).

The grating equation predicts that the diffraction of the incident wave front can be derived from the analysis of the condition that guarantees constructive interference of the waves.

To obtain constructive interference, the path difference between incident and diffractive wave fronts must be some multiple of the spatial period of the radiation, i.e. the wavelength, λ.

As described in Figure 2.17, the path difference between rays from adjacent grating centres is given by $(a - b)$.

Using simple trigonometry, this can be transformed into:

$$d \times \left[\sin(\alpha) - \sin(\beta) \right]$$

where d is the grating period (angles are positive when measured anticlockwise relative to the grating normal).

Ensuring that the interference is constructive, we obtain the equation:

$$m \times \lambda = d \times \left[\sin(\alpha) - \sin(\beta) \right]$$

where m is an integer (called the diffraction order).

The performance of a simple surface-relief diffraction grating for normal incidence can be obtained from the previous general case, by making the incident angle zero, thus obtaining the equation:

$$m\lambda = d \sin \theta$$

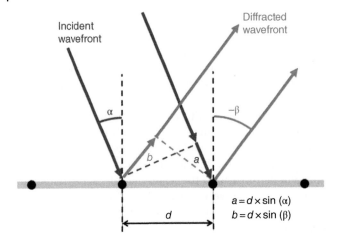

Figure 2.17 Calculating the path difference between adjacent grating centres. Figure courtesy of Alex Cabral.

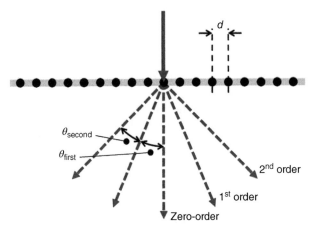

Figure 2.18 Orders of diffraction for a transmission grating. Figure courtesy of Alex Cabral.

This relationship defines the incidence of light of wavelength λ falling perpendicular upon a surface grating of line spacing d, where the order of diffraction is m and the angle at which that order is diffracted is θ.

Thus, for the zero order (that is, when $m = 0$), the "*undiffracted*" beam passes directly through the grating, as shown in Figure 2.18.

In holography, we are predominantly interested in first order diffraction. If the incident beam arrives at an angle to the surface, the relationship of the grating equation still holds for the first order diffraction:

$$\lambda = d\left(\sin\theta_{\text{dif}} - \sin\theta_{\text{i}}\right)$$

where λ is the wavelength, d is again the *period* of the grating, or the distance between its fringes or lines on the surface of the material, θ_{i} is the angle of incidence at the grating and θ_{dif} is the angle of the diffracted beam from the normal.

It was Lord Rayleigh who first suggested that the zero order light, which is regarded as "wasted" in such applications, could be reduced by controlling the profile of the ruled structure, and this inspired Robert Williams Wood to produce the first "blazed gratings" in 1910.

But in a layer of infinitesimal thickness, where only the lateral spacing of the fringes is relevant, the grating equation above predicts that two first order beams will appear on either side of the zero order beam, as we have seen.

In practice, in the recording of holograms in silver halide, DCG or photopolymer, the second of these diffracted elements is, in fact, very dim; so our holograms do not appear, at first sight, to obey the grating equation in its simplest form.

In a thick hologram, which is defined as a hologram where the *thickness of the recording layer is significantly more than the wavelength of the light involved*, the Bragg condition assumes great importance; it is the explanation for the exceptional diffraction efficiency that we can achieve when recording in these "thick" materials.

2.18   Bragg's Law

Early in the twentieth century, William Lawrence Bragg, working with his father, William Henry Bragg, recognised that crystalline materials scattered x-rays in predictable peaks, which they showed to be a function of the inter-atomic spacing of the crystal. X-ray diffraction scanning eventually became a classical method for the analysis of crystal structure and was used in my own experience for silver halide crystal analysis, as previously mentioned.

The Braggs showed that the relatively simple equation:

$$n\lambda = 2\,d\sin\theta_{\text{Bragg}}$$

was able to predict the precise spacing between crystal planes.

Despite the apparent discrepancy in the formulae for Bragg diffraction and the general grating equation, this Bragg equation *is* in accord with the general case discussed previously, as the angle of incidence is equal to *minus* the diffraction angle. We suffix the angle symbol, θ, here with the subscript ($_{\text{Bragg}}$) to delineate the use of the common angular symbol between these specific cases.

Figure 2.19 shows that the angle of incidence of the x-ray beam will dictate that the wavelength will coincide with the atomic spacing in the crystal in discrete peaks when reflections from the various layers are in phase; that is, those reflections interfere constructively.

Here, n is an integer or whole number;
   λ is the wavelength of the x-rays;
   d is the spacing of the crystal planes or atomic centres; and
   θ_{Bragg} is the angle of inclination of the beam incident upon the crystal surface.

Of course, x-rays are a small part of the electromagnetic spectrum and meaningful laws must apply "across the board". The principle of Bragg diffraction is equally meaningful in the visible wavelengths and the Bragg condition has become a vital part of the design process for volume holograms.

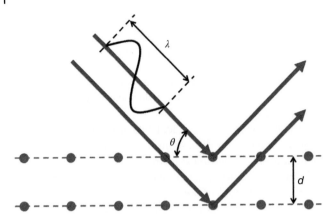

Figure 2.19 The Braggs' x-ray crystallography analysis. Figure courtesy of Alex Cabral.

In the case of holography, we are dealing not with crystal planes, but with the similar construction of layers of refractive index modulation which we create in the production of a bleached phase hologram. This introduces the concept of the *volume hologram*, which automatically defines the thickness of the recording layer in relation to the planar fringe structure, and which concept differs inherently from the *thin hologram* that forms the basis of the surface-relief hologram, nowadays so common as embossed holography in the security industry.

In this volume configuration, these planes of index modulation act as a stack of partially efficient mirrors and mathematically we can show that, because of the specific period of the grating with respect to the wavelength of the particular monochromatic light which will reconstruct the hologram, a wave crossing an individual refractive index interface (which, in the case of silver halide recording material, may arguably be regarded as a sine or cosine variation in refractive index) will arrive at the subsequent interface(s) in phase, and will therefore be reflected in phase, so as to create constructive interference in exactly the same way as the Braggs' experiments with x-rays.

If we arrange for our bleached phase hologram to meet this Bragg condition perfectly, then it can reach a theoretical maximum of 100% diffraction efficiency (D.E.).

The work of Kogelnik and others in this area [5] predicts that there are obvious conditions including level of index modulation and emulsion thickness itself which can easily detract from this maximal theoretical D.E.; but it is true to say that we have often made diffuse image master holograms which cast a very dark shadow indeed in the position where zero order laser light would be expected to fall on the wall of the studio!

If two plane wave fronts arrive at the surface of a recording medium, as shown in Figure 2.20, the planes of interference bisect the angle between the wave fronts.

It has frequently been asked, with regard to volume holography, whether the effects of Snell's Law will modify the grating microstructure due to angular changes when light enters the glass recording plate.

As the refractive index of most common materials at the x-ray frequencies of the electromagnetic spectrum is close to unity, this had no relevance in calculating the plane spacing in crystallography in the original x-ray crystallography work by Bragg and son, when originally deriving the Bragg equation. (The reason for the use of x-ray

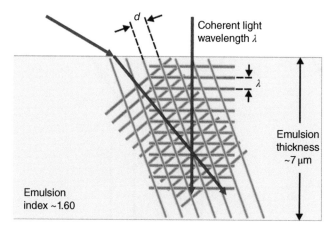

Figure 2.20 Fringes bisect the angle between incident beams.

diffraction in crystallography is because its wavelength between 0.1 and 1.0 nm, as was shown in Figure 2.4, coincides precisely with the typical order of dimension of crystal planes.)

Equally, when applied to thin transmission holograms, calculation of the grating spacing is again influenced by the unity value of this parameter. The surface-relief recording materials, such as photoresist and especially photo-thermoplastic film, are, by definition, effectively close to zero thickness. In terms of fringe dimensions, for such transmission holograms, the fringe structure is effectively defined in air.

But the practitioner of the volume hologram should not be concerned by this phenomenon. When we consider the passage of light through a medium of higher refractive index than air, we are aware that frequency, the time-related parameter, cannot change (conservation of energy), and we know that temporal coherence is preserved, so it is therefore the case that the wavelength must be considered to vary from the vacuum value during its transit through the higher-index medium.

As mentioned in Section 2.7, the speed in a medium, for electromagnetic waves, is governed by the refractive index, n, of the medium ($s = c/n$).

Therefore, the wavelength of the radiation in a medium with refractive index n is given by:

$$\lambda_{medium} = \lambda_{vacuum}/n$$

When one wants to determine the fringe spacing of two-beam interference in vacuum (or air, since for holography this approach is valid, as mentioned before) with the grating equation, in the configuration shown in Figure 2.21(a), we obtain:

$$d = \lambda/\left(2\sin\theta\right)$$

In the case of the same two interfering beams (with an angle 2θ between them) entering a medium with refractive index n, the angle is now ε, and it can be obtained from the Snell equation $\sin(\varepsilon) = \sin(\theta)/n$.

(a) (b)

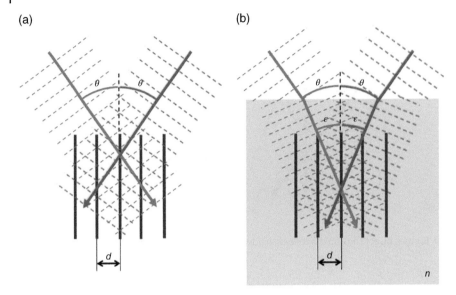

Figure 2.21 (a) Fringe spacing as calculated in air; (b) fringe spacing as calculated in glass. Figure courtesy of Alex Cabral.

So the spacing derived from the grating equation for the interference inside the medium in Figure 2.21(b) with refractive index n for the radiation that now effectively has a wavelength of $\lambda_{medium} = \lambda/n$ is given by:

$$d = \lambda / n / \left(2\sin\varepsilon\right)$$
$$= \lambda / n / \left(2\sin(\theta)/n\right)$$
$$= \lambda / \left(2\sin(\theta)\right)$$

Thus, we have shown that the fringe spacing inside the medium is the same as if it was in vacuum. For simplicity, all the calculation that we need to do can be done without considering the change in angle due to refraction, provided we also do not consider the change in the radiation wavelength.

In calculations involving fringe shrinkage etc., relating to thick holograms, the refractive index values of silver halide and photopolymer are taken into account, and, in fact, the advanced photorefractive crystal applications of holography are literally a function of the ability of electrical charge to modulate refractive index; thus, index inclusion is fundamental in such specialist applications of volume holography. If it is the intention to modify fringe structures, for example, chemically or electrostatically, it is, of course, necessary to consider a precise account of the configuration of the microstructure.

2.19 The Bragg Equation for the Recording of a Volume Hologram

As we have described, the Bragg equation considers the space between the fringes *within the layer* rather than the spacing at the surface of the film, which we have encountered in the grating equation.

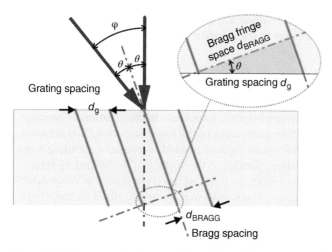

Figure 2.22 Bragg spacing between fringe planes. Figure courtesy of Alex Cabral.

From Figure 2.22, with angle φ between two incident beams, the relationship between the grating spacing at the surface, d_g, and the perpendicular spacing between the planar fringes, d_{Bragg}, is:

$$\cos\theta = d_{Bragg}/d_g$$
$$d_{Bragg} = d_g \cos\theta$$
$$d_g = d_{Bragg}/\cos\theta$$

Therefore, the basic grating equation, previously defined as:

$$\lambda = d_g \sin\varphi$$

for first order diffraction, where φ is the angle between the beams, then takes the form:

$$\lambda = d_{Bragg} / \cos\theta * \sin\varphi$$

In this case, the angle between the beams $\varphi = 2\theta$.
But in trigonometry, $\sin 2\theta = 2\sin\theta . \cos\theta$. So,

$$\lambda = \left(d_{Bragg} / \cos\theta\right) * 2\sin\theta . \cos\theta$$
$$\lambda = \left(d_{Bragg}\right) * 2\sin\theta$$
$$\lambda = 2\, d_{Bragg} \sin\theta$$

which is the Bragg equation.

Note: Bragg defined "θ" as the angle between the crystal planes and the incident ray – in this case, half the angle between the beams.

At the recording stage, as shown in Figure 2.22, the two incident waves arrive at the silver halide recording material from the same side of the recording material. The general level of inclination of the fringes recorded in the film can be influenced by changing the angle of incidence of one or both of the beams. But, as in this example, typically for a display hologram, one beam is normal to the surface of the recording material.

(Generally in display holography, we must always consider it likely that the "object beam" is preferred to be perpendicular to the recording plate.)

2.20   The Bragg Condition in Lippmann Holograms

We have shown that the Bragg condition is vitally important in the creation of "volume" holograms. As a medium for reflection holograms of the Lippmann type, this phenomenon is peerless. Here, the thick hologram volume is used to record a standing wave whose planar fringes lie, to some extent, parallel to the surface of the recording layer.

In this case, the modulation of the refractive index of the layer results in the production of a reflective filter which is *extremely* sensitive to the wavelength of incident light, such that, for a given angle of incidence, only a very narrow band of wavelengths will be reflected. This is the reason why such a hologram is able to be viewed in incoherent white light – *the white light reflection hologram.*

This part of our holographic technology resembles the principle by which "dielectric high efficiency" mirror coatings have been made for many years by the deposition of alternating quarter-wave thicknesses of materials of successively high and low refractive index. Note also, at this moment, that the familiar production of anti-reflection coatings in areas such as camera or spectacle lenses is effectively a result of the deliberate *calculated abuse* of the Bragg condition under discussion here!

However, the ability in reflection holography to modulate the refractive index in an image-wise fashion, or in a fashion where *the layers are other than parallel to the substrate surface,* is a really powerful holography technique, which defies the deposition technology.

Reconstructing the hologram from the reference angle, as shown in Figure 2.23, we see that constructive interference will occur only when the appropriate wavelength is incident; other colours will pass through the volume grating without effect, so the *reflected object beam* in the diagram is monochromatic, despite the white illuminating beam.

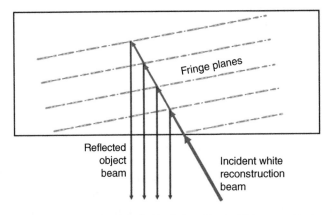

Figure 2.23   **Principle of the "white light reflection" (Lippmann) hologram.**

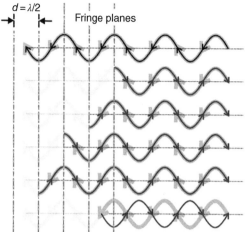

$d = \lambda/2$

Fringe planes

Figure 2.24 Summation of in-phase reflection. Figure courtesy of Alex Cabral.

The requirement for the Bragg condition is that the individual fringes, or index-modulated interfaces, in the recording layer lie at such a frequency that a component wave front passing the first plane without reflection will arrive at the second layer and subsequent partially reflective surfaces *in phase*. Its reflection will then travel in the same direction, in the same phase, as the light reflecting from the first fringe plane, and will thus interfere constructively with that light. If the index modulation is sufficiently high and if there are sufficient layers of modulation in the hologram, light of a single wavelength may eventually be reflected in summation to 100% of the incident beam.

The condition for this constructive interference is clearly that the wave advancing beyond the first fringe layer will reflect from a subsequent layer to arrive in phase with light from the first layer. In Figure 2.24, the wave front in the lowest of the wave profiles shown reflects from an interstitial index modulation and is thus reflected out of phase, resulting in destructive interference; this being equivalent to the effect of incident light having some other wavelength.

In the very simplest case, in accordance with Bragg's Law, we can say that when two beams from opposite sides of the emulsion are coaxial, the angle between these beams is 180°, and thus $\theta/2$ is 90°. As sin 90° = 1, we then have the condition in the Bragg equation, where the fringe spacing is half the wavelength:

$$\lambda = 2d \sin(\theta/2)$$
$$\lambda = 2d * 1$$
$$d = \lambda/2$$

The condition for the constructive occurrence, as predicted by Bragg, is that its path length from the first layer to the second layer, plus its return to meet light reflected from the first layer, is one wavelength (strictly "an integer"). This effect is shown in Figure 2.25 in the more practical situation where the reference beam is off-axis, in the style of Denisyuk, indicating the phase condition of the wave fronts of incident and reflected light as they interact with the planar grating.

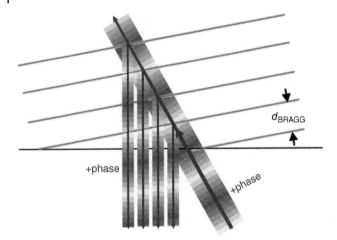

+phase

+phase

d_{BRAGG}

Figure 2.25 Off-axis (Denisyuk) phase preservation geometry. Figure courtesy of Alex Cabral.

2.21 The Practical Preparation of Holograms

Provided a technician has a basic command of the physical principles in this chapter, there is nothing to stop them reliably planning and producing successful holograms.

The holography lab is a very good place to achieve an understanding of basic optics, because provided the correct health and safety measures are in place, the use of visible light lasers enables the technician to gain an exciting and clear visual experience of the relevant laws of optics which have been detailed in this chapter.

It is also to be noted, however, that the production of a really successful display hologram demands an unusual combination of skills in Physics, Chemistry and Art & Design, as well as great dexterity and patience!

Notes

1 Hecht, E. (2015) *Optics*, 5th edition. Pearson (ISBN: 978-0133977226).
2 Sony Music.
3 Sambles, J.R. and Lawrence, C.R. (2002) Limited-view iridescence in the butterfly *Ancyluris meliboeus. Proceedings of the Royal Society B: Biological Sciences*, 269(7), 7–14.
4 Figure 2.12 L2012 Pool Camera reproduced by kind permission of IOC.
5 Solymar, L. and Cooke, D.J. (1981) *Volume Holography and Volume Gratings*. Academic Press Inc. (ISBN: 0-12-654580-4).

3

Conventional Holography and Lasers

3.1 Historical Aspect

We have seen how Lippmann set a precedent for imaging science when he recorded colour photographs based upon interference with natural white light as his power source.

But even Denis Gabor did not have the essential working tool, the laser, [1] when he invented holography. (Acronym: L.A.S.E.R. derived from Light Amplification by Stimulated Emission of Radiation. Due to the public perception of "death rays" etc., it is not always appreciated that the word is an acronym!)

Gabor used a mercury arc lamp to provide a partially coherent source of light for his work. In hindsight, with our ability now to overview the problems experienced when our lasers malfunction, *with sincere self-pity*, we can see exactly how difficult this would have made the pioneering task of inventing a whole new imaging system in the absence of bona fide coherent light sources. So the admiration for Gabor's work grows on a daily basis as we work with our modern equipment.

As shown in Figure 3.1, the mercury arc contains a strong line at 435.8 nm which, as mentioned in Chapter 5, is the explanation for the chosen sensitivity peak of the Shipley photoresist described later, which has so conveniently allowed the 442 nm helium–cadmium and the 458 nm argon laser lines to progress embossed holography at such a phenomenal rate into the form of a billion-dollar industry; the photoresist was by no means originally designed for holography, and our relatively minor take-up as an industry could not justify large research budgets from the manufacturer.

Today, we have a wide range of laser devices and the marketplace is changing on a monthly basis as the technology develops at an incredible rate. If we consider that "any choice is better than none", then, with hindsight in regard to the difficulties experienced by the pioneers previously mentioned, there is, in fact, absolutely no excuse for error in our choice of lasers from the available range in this day and age.

In the 1980s at Applied Holographics, the late, great Professor Nick Phillips discussed with us the idea that, with special attention to technique, we could perhaps use a sodium vapour lamp, with its glorious orange–yellow output of semi-coherent light, *the sodium D-lines* (Figure 3.2), to produce certain contact holograms.

After much discussion, though, we observed that life is short, and with that in mind, we concluded that it would probably be better to concentrate on creating really productive

The Hologram: Principles and Techniques, First Edition. Martin J. Richardson and John D. Wiltshire.
© 2018 John Wiley and Sons Ltd. Published 2018 by John Wiley & Sons Ltd.
Companion website: www.wiley.com/go/richardson/holograms

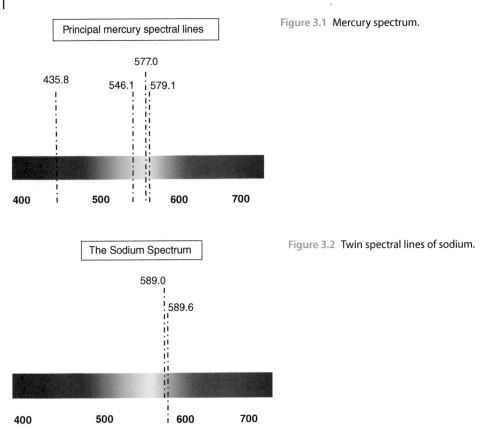

Figure 3.1 Mercury spectrum.

Figure 3.2 Twin spectral lines of sodium.

methods of using the 100 mW helium–neon laser on the optical table before us, with its half-metre coherence length, to produce better holograms.

Since that occasion, and especially since the commercial advent of the diode laser, I have had numerous encounters and discussions with technicians who have acquired a laser that was made for the purpose of light shows etc., and, with *assurance of SLM mode of operation*, have embarked upon a holographic voyage of frustration.

Recording materials, chemicals and working time are expensive commodities, and the stated empirical nature of this book brings a responsibility to point out that *carefully chosen lasers are the most fundamental requirement for the successful holography studio.*

3.2   Choosing a Laser for Holography

One of the problems faced by a newcomer to the simple conventional holography process is that the expectation to create a working recording system effortlessly can be thwarted by the wide range of minor features of an optical system, or the recording material and process regime, which, more often than not in our experience, prevent immediate success.

In our opinion, the apparently straightforward technical description of the theoretical principle of recording a hologram, in many advanced publications, frequently conceals potential fatal practical difficulties which can bring the debutant "hands-on" technician to a state of extreme frustration. As a beginner to the technique, the various issues of laser stability, lack of coherence, component movement, air flow, chemistry, etc. can produce very similar defects in a recording, which all effectively amount to unspecified failure.

This might easily be the key to the longevity of holography in the security sector, because there are very few people in the world capable of navigating the stormy seas which protect the multi-disciplinary methodology involved in the production of world class security holograms. Furthermore, the major players, such as Opsec, have long since recognised the advantage of combining techniques in a single security hologram in order to challenge the counterfeiter to master more than one area of the technology.

There are endless horror stories regarding expensive lasers delivered for holography that proved to be unsuitable, or at least extremely limiting for the holography system. We would suggest that the fundamental first requirement for any real success in the laboratory is a suitable and reliable laser. This is the *engine room* of the holography laboratory and it should be treated with the utmost respect – a relationship that, unfortunately and necessarily, always stops short of complete trust!

During our careers, we have seen incredible advances in laser technology, but the growth of the laser industry now leaves us, to some extent, spoiled for choice, to the point where the selection of the *most appropriate* laser remains the key step in setting up a successful laboratory; but, in fact, this expanding selection process simultaneously provides an *ever-increasing* opportunity for expensive mistakes.

There is every reason, therefore, to flag up the pitfalls which exist in laser acquisition: there are so many apparently attractive lasers to be acquired nowadays, but we frequently hear tales of lasers which do not lend themselves ideally to general holography work.

Twenty years ago, the purchase of a gas laser was such a major event, as regards the cost and provision of power and cooling infrastructure, that it became an issue of massive significance to any company. However, advances in diode and laser technology mean that realistic levels of power and coherence can be achieved at costs reduced by perhaps an order of magnitude; with this freedom comes an opportunity for mistakes!

Every holographer cannot necessarily be an expert in every aspect of the growing wealth of laser technology. But we need a basic knowledge of how and why the various types of laser function; so we describe below the basic history and principles of the process of creating monochromatic coherent light. Laser *power* is a prime consideration for holography. In Chapter 5 we describe the recording materials that are available for recording holograms and we can estimate their sensitivity. But the range of sensitivity between a silver halide, a photopolymer, a photoresist or dichromated gelatin probably spans up to two orders of magnitude.

We also need to take into account the proposed dimensions of the recording medium as well as much more complicated power requirement assessments, such as the adequacy of the laser to light the particular subject matter, which, in turn, may be a "lossy" solid object of unspecified reflectivity, or a backlit screen, which typically could be illuminated via a spatial light modulator. In many cases, the whole system of object illumination may be quite inefficient in terms of light arriving at the master hologram.

In general, a more powerful laser will facilitate shorter exposures and will thus be less demanding of the stability requirements of the optical table; but it may also bring a demand for high-power optics and increased health and safety issues.

3.3   Testing a Candidate Laser

Above and beyond all other considerations, it is a really wise precaution to arrange a test session prior to purchasing a laser; most distributors recognise the rather mysterious nature of producing hologram masters, and in many cases will genuinely welcome a test session to provide the confirmation that their laser is capable of such work in the uncharted waters of holography, especially if you are prepared to supply a technical report on the advantages and limitations of the laser for your purpose.

Now that water cooling is generally unnecessary, the logistics of a testing session are relatively simple.

Later, in Chapter 8, where we show example layouts for working optical tables, we will recommend set-ups which will test the usefulness of the laser for individual needs of holography systems, without the need for too many optical components. One of these is the Michelson interferometer, which is the standard recommendation to the budding holographer due to the sheer technical elegance of the technique; but we also provide an alternative idea.

The Michelson interferometer was originally designed to delve into "Special Relativity" and to investigate the suggested "drift of the aether". Albert A. Michelson and Edward Morley explained, in 1887, a method where light beams from a single source were organised to travel orthogonally before being reflected back to a central position where their interference pattern was studied. Thus, if there was a principally linear movement of the containing envelope of "space", one of these light beams might be travelling with apparently modulated velocity. As far as the experimenters were concerned, the results were negative then, but, with the advent of the laser, the method can now offer a means for measuring positional stability, vibration and, importantly, laser coherence.

Recent news headlines have heralded the experiments confirming the existence of "gravitational waves" and the Michelson–Morley interferometer principle appears to be the centrepiece of this experimentation, almost without general acknowledgement.

Vibration can even be categorised as regards its sources and localised influences such as doors, walkways, etc., with a visual display of their effects. We discuss this fully in Chapter 7.

As an experienced holographer, confident of controlling other potentially confusing properties of the system, when solely testing a prospective new or replacement laser, we should not overlook the fact that there is sometimes a great deal to be said for building a simple Denisyuk holography testing camera, because it may take a minimum of major optical components; if this works, we have a *direct* conclusion regarding the ability of the table to make holograms, rather than interpreting how the interferometry results impact upon the ability to do so!

Additionally, the method of making this simple but demanding type of hologram automatically answers the question of whether the laser is of sufficient power to make holograms of the selected dimensions without introducing calculations which necessarily entail theoretical assessments of film sensitivity, Gaussian distribution of the laser beam and its effect on gamma and density of the recording medium.

One prominent difficulty with the use of lasers which have not been designed specifically for holography is the contradiction between the need for cooling systems of any type and the absence of vibration. The currents of air caused by convection or powered cooling fans can be problematic in the extreme, and laser designers are not always sympathetic to the requirements of holography.

Although the problems associated with water cooling of gas lasers appear to be behind us, some of the new d.p.s.s. lasers acquired recently include cooling fans which produce far more noise and air currents than one would hope.

3.4   The Race for the Laser

In 1957, there was a general awakening that light amplification was feasible and the race for the invention of the laser began in earnest. In the USA, it was anticipated that military applications would follow. Charles Townes and Arthur Schawlow (Figure 3.3), who were brothers-in-law working at Bell Labs, had previously mastered the parallel concept for microwave radiation (MASER), and continued to term the target product the "Optical Maser."

They took part in an exciting and intensely competitive experimental period of research with the aim of demonstrating the proposed new technique. A diverse range of gaseous and solid lasing media was considered, but it was, in fact, the ruby system which finally delivered the world's first flash of laser light.

Townes and Schawlow were effectively in direct competition with Gordon Gould at TRG, and indeed it was Gould who was actually using the term "LASER" whilst Townes continued to refer to the "Optical Maser". But at TRG, Gould was working under a classified contract, because the US government recognised the possible military value of the laser, which led to delays as a result of astonishing political problems; as if the difficulties of the technology itself were not sufficient!

Figure 3.3 Pioneer brothers-in-law.

As Gould had previous connections with communist ideals, he was unable to gain security clearance to work on the project with its apparent military applications, and the progress of his Technical Research Group was thus retarded enough to lose the race for the prize of giving the world's first demonstration of "laser" light. Foiled by politics!

Theodore (Ted) Maiman, at Hughes Aircraft Laboratories, finally stole the show with the first successful demonstration of the ruby pulse laser on 16th May 1960, but the battle for patents continued thereafter, in relation to Gould's earlier ratified notes, and the legal battle went on for thirty years!

Jeff Hecht's excellent book *Beam* [2] brilliantly captures the historical drama and some useful technical detail of the experiments which yielded the first lasers for those who are interested in this aspect of holography.

Having worked intensively with JK Lasers in the 80s during the development of an exciting new generation of commercial ruby pulse lasers for holography, the authors can imagine the sheer excitement of seeing the first flashes of pure red light which were emitted in Maiman's laboratory to change history on that momentous occasion.

3.5 Light Amplification by Stimulated Emission of Radiation (LASER)

In simple principle, the configuration of the laser is a cavity in an active medium with reflective surfaces at its ends, along the lines of the Fabry–Perot etalon.

Figure 3.4 shows a schematic of the basic laser configuration which can be adapted to suit solid and gaseous lasing media. Suitable light is pumped into the cavity from the sides. So a ruby rod is an ideal medium for the purpose.

The pump light raises electrons in the lasing medium to a high energy level, and if the pump light is sufficiently intense, a *population inversion* is reached, whereby we have more atoms in the excited state than in the ground state. This phenomenon of creating a population inversion is the vital step in producing laser light, and explains why the study of the intermediate energy levels in the available orbitals of various atoms has been able to predict the candidate materials for successful lasing.

Fortuitously, particle physics defines that the collision of photons with suitably excited atoms results in an emission which is *related in phase* to the incident photon. Therefore, we have the condition for stimulated emission, since coherent waves are thus produced by the lasing action.

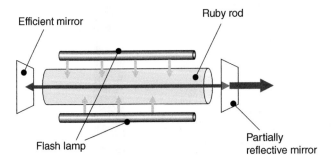

Figure 3.4 Ruby laser schematic.

3.6   The Ruby Laser

Ruby is, in fact, crystalline aluminium oxide, doped in various degrees with chromium. The chromium ion absorbs light of blue and green wavelengths and thus colours the crystal pink. The strongly coloured ruby which is the most desirable natural material for jewellery is not the preferred material for lasers; a pale pink colouration is preferred for laser construction, but nevertheless the rods used in modern pulse lasers have an awesome beauty.

There is no requirement to coat the sides of the pale pink artificial ruby rod, and thus energy is easily entered naturally into the system from the sides of the cylindrical crystal. Much of the early experimentation concentrated on the difficulties of producing sufficient light to pump the lasing medium. In fact, at the time of Maiman's work, the elusive pump source was the most critical component and a helical flash tube was used for the first successful laser.

The end mirrors are adjusted to ensure a cyclic reflection of light through the rod; one mirror is of the order of 100% efficiency whilst the other allows a small fraction of the light incident upon it to exit the rod as a laser beam. In the early days, the ends of the rod itself were polished perfectly parallel and coated with reflective metal; nowadays, in practice, the polished ends could be coated with an *anti-reflective* layer to avoid spurious reflections, whilst separate mirrors, which can be aligned with absolute precision via adjustable mounts fixed very firmly to the laser rod housing, are used to form the lasing cavity. A material such as *Invar* is able to provide exceptional dimensional stability and insensitivity to temperature change. There is great skill involved in "walking-in" the reflective surfaces to achieve perfect alignment. With a pulse laser, we do not have the advantage of real-time visual assessments of beam quality during the adjustment process. My colleague and great friend, the late Rob Rattray, became expert at this process whilst dealing with the JK ruby lasers at Applied Holographics.

Because the profile of a 15 nanosecond ruby flash is visually imperceptible, we use photographic paper fogged and developed to medium density as an observation screen. The violent incidence of the pulse beam will spontaneously bleach this paper in a single pulse and provide a good record of the beam profile. As the "walking-in" process occurs, the beam recordings improve. Obviously, to visualise the beam in real time is a necessity, so this method provides an excellent solution. A "burn" sequence of this type is shown in Figure 3.5.

In practice, the modern pulse ruby laser comprises the system described above in the form of an "oscillator" which is used to pump a second ruby which has been excited but whose ends are non-reflecting. This rod acts as an amplifier to the original pulse and the whole unit is thus capable of exceptional power.

The JK Lasers HS10 Ruby pulse laser, with which the authors made many well-known holograms illustrated in this book, had the ability to concentrate up to 10 Joules of energy into a time interval ("pulse width") of 15 nanoseconds. Without the Q-switch, the pulse is extended into the order of milliseconds in accordance with the flash conditions.

In this general configuration, a Q-switch (Pockels cell) is included *within* the modern pulse laser cavity (i.e. between the mirrors). This electro-optic crystal has the ability to rotate the plane of polarisation in the same way that a wave plate operates, but is electrically driven. Operating within the cavity of the modern pulse laser, it has the ability to

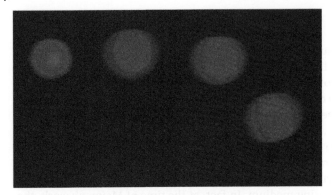

Figure 3.5 "Burn-paper" sequence of ruby laser tuning.

concentrate a large portion of the flash tube energy (millisecond timespan) into a single pulse of the order of 15 ns.

With the Q-switch (so called in relation to the Quality factor for the gain medium) functional, the pump light is able to create a population inversion, the highest energy state within the ruby, before the lasing action begins, so as to create a giant pulse of laser light over a period of time of only 10–20 ns. Just to understand the magnitude of this pulse, we can consider the power in ordinary terms *whilst the laser is "on"*.

$$10\,J\,in\,15\,ns = 10/\left(15 \times 10^{-9}\right)$$
$$= 0.6 \times 10^{9}$$
$$= 600,000,000\,Watts$$

Fortunately, the laser cannot "stay on"! The giant capacitor bank has been emptied entirely by each flash of this magnitude, and will take several seconds to replenish. But the enormity of this instantaneous "power" assessment reminds us that we are in a new realm in terms of eye safety.

During this incredibly short time period of the laser pulse, most subject matter can be regarded as completely stationary, so that the traditional holographic "movement" gremlin is automatically eliminated. This, of course, opens the door to the use of pulse lasers for human portraiture and other artistic use of holography to record three-dimensional scenes from life. Non-rigid models can be used and any number of ordinary objects. My co-author has utilised this advantage of pulse lasers in a career stretching from the installation of a JK Ruby pulse laser at the Royal College of Art in 1984 to the present day, and a number of very impressive pieces have originated from this technique. First generation masters made using the stability advantages of a 15 ns pulse ruby laser light can be transferred into rainbow transmission or reflection holograms as a second generation copy. An example of a ruby origination which met with commercial success is the hologram photographed in black and white in Figure 3.6, entitled "The Mathematical Chef".

However, the pulse length is so short that we have a very special requirement for silver halide films to be absolutely free of high-intensity reciprocity failure (HIRF). The reaction of a recording medium to this light is also capable of being quite different to the typical exposure of the material to "ordinary light" because the time span involved is not dissimilar to that encompassing molecular events in a chemical reaction – we will mention this in Chapter 5.

In practice, this dictates that certain sensitising dyes are not suitable; the films containing them are not capable of pulse laser work – it is as if the energy is gone before the necessary chemical changes occur.

One of the exceptional ironies of successfully operating a ruby laser is that the rather low optimal temperature for lasing of the system conflicts rather seriously with dew point considerations in summer conditions in Britain; there is a narrow window of operation unless humidity control measures are undertaken, because the ruby optimal lasing temperature tends to coincide with the dew point in summer.

Users of conventional c. w. lasers are often surprised to find that the output beam of a typical pulse ruby laser is of such a large diameter, directly related to the diameter of the amplifier rod, for example of the order of 10 mm diameter or more.

Figure 3.6 "The Mathematical Chef" – self-portrait of Martin Richardson.

3.7   Laser Beam Quality

There was an interesting race to provide spatial filtration of the pulse ruby beam in the 1980s. The output beam of the laser was typically slightly divergent due to the tendency for collimation to result in the production of a "donut" profile in the far field. Negative lenses are used to diverge the beam for use, since the act of focussing such intense energy will create difficulties in air, involving explosions capable of creating phase conjugate reflection, which may ultimately damage the laser.

As regards spatial filtration, Figure 3.7 shows a silver halide recording of a spread beam from the JK/Lumonics HS10 Ruby laser without any filtration. We can see that the distribution of light itself is relatively homogeneous; most of the disturbance we see here is clearly due to internal reflection in the recording film at normal incidence, and hence the beam quality is well suited to recording phase holograms of clean appearance. The photograph in Figure 3.8 shows that there is probably less homogeneity associated with a typical helium–neon beam without filtration.

Therefore, our experience generally shows that spatial filtration is certainly not a necessity for the recording of redundant master holograms and barely advantageous in the contact copying mode, as proven in the Applied Holographics pulse ruby HoloCopier;

Figure 3.7 Ruby beam without filtration.

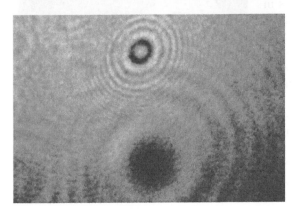

Figure 3.8 The HeNe equivalent.

in practice, the difficulty of keeping up-stream optics perfectly clean is far more relevant than filtration of the source beam.

Dust that gathers on any mirrors will cause major difficulties, especially in the areas where the beam energy is concentrated prior to divergence. In fact, at such peak energy levels, such particles will typically fuse into the mirror surface to create permanent damage.

The subjective experience of exposure to such a short pulse of light is a strange phenomenon, arguably surreal; it is as if the senses can detect that the light is gone almost before our perception that it is present kicks in.

Our practical experience requires us to point out that the response of the human eye is *extremely* limited at the wavelength 694 nm of the ruby laser. When we see spurious ruby light that has eluded a masking system on the optical table fall on the studio wall, the subjective impression is of a low level of illumination perceived by the eye. But the reader will see from the plot in Figure 3.9 why it was the case that the authors, for so many years familiar with that sight in the corner of the eye of excess ruby light falling on the distant studio wall, were truly shocked by the remarkable brightness in similar circumstances of accidental exposure to a 532 nm laser Nd:YAG pulse, of equivalent power, in our earliest experiment with two-colour pulse holograms.

This is a very serious matter which the reader must take into account to avoid eye injury – there is no second chance with a pulse laser!

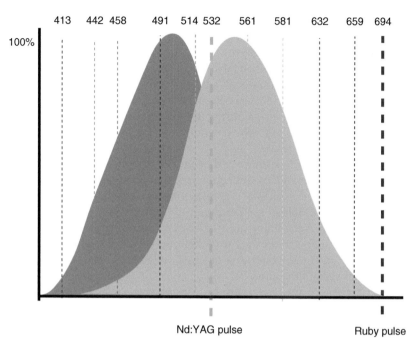

Figure 3.9 Pulse laser wavelengths as part of the spectrum of human vision.

3.8 Photopic and Scotopic Response of the Human Eye

The sensitivity of the human eye is conveniently categorised into two forms corresponding to high intensity (daytime) vision, called *photopic vision,* and low intensity (night) vision called *scotopic vision.*

Of course, there is a wide range of variability in an individual's perception of light, which varies with age and must include features of colour blindness, which amount to a shift of wavelength sensitivity that may entirely exclude longer wavelengths, but ordinarily the peak sensitivity occurs at a wavelength of 555 nm (yellow-green).

Scotopic vision relates to low light levels, and in this case the sensitivity of the eye is mediated by rods, not cones, and shifts toward the shorter wavelengths, peaking at around 507 nm for young eyes (blue-green).

Later, we will further consider the response of the human eye with respect to the design of reflection holograms and the selection of lasers. Figure 3.9 shows the approximate relative eye sensitivity in relation to the common wavelengths which are available through stimulated emission. It would be careless to ignore these concrete facts when designing holographic images.

3.9 Eye Safety I

The question of eye safety is extremely simple when making holograms by pulse laser, because there is simply no feasible advantage in any attempt to visualise the laser beam. For this reason, the widespread use of safety goggles or glasses is unquestionably necessary and provides no inconvenience to the technician.

The health and safety eye protection issue is far more complicated when continuous-wave lasers are used in the business of making holograms which are for visual appreciation of the eventual viewer, in both the display and security areas.

It is true that digital holograms can be made without routine visual involvement of the technician with laser light; but the conventional "analogue" type of hologram necessarily involves the technician in viewing certain laser-lit scenes.

Recently, the advent of new techniques utilising the newly available Class 3B (essentially < 500 mW) and Class 4 lasers (>500 mW) in a range of wavelengths of emission has further complicated the eye safety issue.

The Class 4 lasers are capable of causing eye damage not only from direct exposure to the original beam but also from diffuse reflections from other surfaces. To the experienced holographer, there is necessarily a real difference in behaviour when using a laser which is deemed to be capable of eye damage from diffuse reflections. Formerly, in the absence of such power in the holography laboratory, for example in the process of aligning an optical system powered by a 30 mW helium–neon laser, where so many holographers began their career, it would have seemed perfectly normal to use a card screen to trace the beam around the table.

However, high-power lasers are readily available nowadays, so if, for example, a 500 mW d.p.s.s. laser of 532 nm falls upon such a screen, there is a real possibility of eye damage from the diffuse reflection. There is also the inevitable conclusion that the card will quickly burst into flames!

You must seek advice regarding eye protection from professional consultants such as LaserMet [3] before using such lasers.

3.10   The Helium–Neon Laser

Demonstration of the helium–neon laser followed rapidly after Maiman's initial success with the ruby pulse. The helium–neon laser was capable of "continuous-wave" operation. Both continuous-wave and pulse, and the obvious hybrid of repetitive pulses, are now useful for many applications in holography.

The helium–neon laser was ubiquitous in optics laboratories in the late twentieth century. Its configuration follows the same principles as the ruby laser wherein a Fabry–Perot-type cavity straddles a lasing medium. Of course, in this case the medium is a gas, so that fresh problems of enclosing this gas at very low pressure whilst permitting the introduction of electrical charge exist.

Inert gases do not involve themselves in the creation of any molecular bonding for obvious reasons; they exist as a mixture of independent atoms. The lasing medium itself in this case is actually the neon gas, which is the minority component (~10%). However, the helium gas is much more than a diluent. It plays a significant role in absorbing electrical energy for a population inversion, in which its electrons are raised to two separate energy levels, which, importantly, happen to be very close to the two upper levels of the neon atom. When the atoms collide in the ordinary course of transit in the cavity, this energy is transferred to the neon atoms. Its exchange at the particular energy level in question results in the release of photons at the well-known 632.8 nm

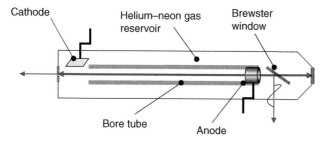

Figure 3.10 Schematic of the helium–neon laser principle.

wavelength, which is the wondrous, familiar scarlet red – arguably the perfect colour for the red component of a holographic RGB tri-stimulus hologram, due to its general aesthetic acceptability and its position on the photopic sensitivity curve previously mentioned in Figure 3.9.

The narrow bore of the laser (plasma) tube provides a restricted volume of gas in which a DC discharge is concentrated, but the 1s excited atoms will tend to return to the ground state due to contact with the sides of the enclosure and, for that reason, unlike the ruby, a narrow lasing capillary is advantageous, as shown in Figure 3.10.

Whereas the ruby excitation is a three-level process, the helium–neon excitation is a four-step energy exchange. The lasing action occurs as a result of transitions between the upper levels of excitement, meaning that population inversion is easy to sustain. This explains the success of the helium–neon laser, whose ubiquitous presence in the physics lab reflects the qualities of:

- Low cost;
- Low power input;
- Air-cooled system – no cooling difficulties requiring water or air movement;
- Reliability;
- Acceptable coherence length (often estimated as half the length of the tube);
- Optimal subjective colour (principally red);
- Extreme durability and longevity (we have experienced lasers which lasted over 25 years with regular use);
- Plane polarisation;
- General convenience and durability of configuration.

The use of "Brewster windows" at the end of the laser with separate cavity mirrors enables us to eliminate light whose electric vector is perpendicular to the mirror plane, since it is substantially reflected away from the axis of the cavity at each pass through the window, whereas the s-polarised light is preserved in the cavity and thus subject to ongoing amplification. In this way, the light from a helium–neon laser with Brewster windows is plane polarised and is thus ideal for holography.

A helium–neon mixture is capable of providing other wavelengths at low power. There are yellow and green ("GreeNe") versions, which were sometimes very useful prior to the advent of the modern d.p.s.s. lasers of multiple wavelengths.

3.11   The Inert Gas Ion Lasers

The argon and krypton ion lasers were the first visible spectrum lasers to provide really high levels of c. w. power to the ordinary physics lab or holography studio. Spectra Physics and Coherent lasers of this type were responsible for the production of many of the world's most famous holograms.

The electron transitions of these inert gases allow the release of photons at a number of useful wavelengths, and the ease of achieving population inversions for the various transitions offers varying output power in a range of useful lines in the classical Innova 90 Laser from Coherent, which has been the bedrock of many successful holography labs. The tube may contain krypton or argon or a mixture of gases. An intra-cavity etalon is able to separate a single line of operation and provide metres of coherence for holographic work.

Of course, because a single tube is able to produce simultaneous wavelength emission, these lasers have found use in lighting displays and there are legendary connections between Nick Phillips's early pioneering holography work and the use of laser lighting by The Who for ground-breaking concerts with laser light shows.

But the holographer is interested principally in narrowing the monochromatic line width to a minimum by the precise temperature control of etalons in order to achieve high coherence lengths.

Coherent's modern lasers [4] featuring "mode-locking" technology in the long-frame Innova lasers make for a minimum of problems for holography when efficient cooling technology is used. The basic configuration utilising argon and krypton gas filling is able to supply a wide range of laser colours at useful levels of output. The most useful of these lines are shown in Table 3.1.

When using high-power lasers, as well as the immediate health and safety issue, another aspect is the need to consider the effect of such energy levels on the optical components, such as mirror coatings, pinholes, shutters, etc., in positions where unspread or focussed beams are directly incident upon delicate surfaces. Such hazards are described in more detail in Chapter 7.

Table 3.1 Inert gas ion laser strong spectral lines.

Argon gas Ar^+	Krypton gas Kr^+	Colour
	647.4 nm	Red
	568.2 nm	Yellow
528.7 nm		Green
514.5 nm		Green-blue
488.0 nm		Blue-green
476.5 nm		Blue
457.9 nm		Mid-blue
	413.0 nm	Violet

3.12   Helium–Cadmium Lasers

During the 1980s when so many small groups were beginning to take interest in the production of master holograms for the rapidly growing Embossed Holography industry, the helium–cadmium laser made its commercial debut. The laser was capable of ultra-violet and blue emission.

At 442 nm wavelength, the HeCd was capable of exposing photoresist materials originally designed for mercury vapour lamp exposure and its light was more actinic than the blue 458 nm argon lasers in mainstream use. (Note that although the argon 488 nm line is more powerful than 458 nm, the spectral sensitivity curve of photoresist negates this advantage.)

But the first commercial HeCd lasers contained a plug of heated cadmium which was mixed isotope, and the resulting coherence length was around 10 cm. This made for great difficulty in setting up a rainbow hologram recording system on the scale generally required for embossed holography. Whilst we struggled at Applied Holographics to accommodate this deficiency by organising a stepped reference beam, it was reported from researchers at that time that single-isotope cadmium would provide a significant increase in coherence.

Laser Lines [5] then presented the Liconix 100 mW Embosser laser, which offered a three-fold increase in coherence length from the same tube. This made embossed hologram master origination a realistic proposition – a process typically involving fairly large silver halide recording plates as well as smaller areas of exposure to photoresist coatings.

The HeCd laser is, in many ways, similar to the HeNe, but temperature control is critical. The air cooling is often supplemented by a remote fan attached to a ventilation trunk which can vent the warm air away from the holography table. The HeCd laser is also able to provide a strong ultra-violet line at 325 nm for other work purposes.

The lasing material, cadmium metal, is, in this case, a solid at room temperature. A plug of solid metal is heated in the reservoir and its vapour is distributed along the bore of the lasing cavity. Careful temperature control can prevent deposition of metal upon the optics and, accordingly, there are detailed specific procedures for closing down and starting the lasers with respect to temperature control.

In common with argon and krypton lasers, the HeCd cavity can be adjusted by the "walking-in" process, which is extremely useful for seasonal temperature adjustment in my experience; adjustments in autumn and spring provide significant improvements in performance.

However, in comparison with helium–neon, the life of the tube of the HeCd tends to be restricted by the deposition of metal in unwanted areas, despite the fact that the procedure of switching on and off is automatically time regulated. Additionally, the cadmium plug is of finite mass and will eventually expire, and the familiar problem of gas leakage, which tends to be the only real weakness of the HeNe system, runs parallel here. As a result, it was rare to achieve more than two thousand hours from each tube in my experience. However, the tube replacement was far less expensive than the argon tubes which HeCd replaced.

Liconix was acquired by Melles Griot [6] in 1998 and the helium–cadmium laser is no longer prominently listed, as attention focusses on diode-pumped solid-state lasers. However, the Kimmon Koha [7] group is still producing its HeCd laser, and far longer tube life is now claimed.

3.13 Diode-pumped Solid-state Lasers

The well-known Coherent Verdi° diode-pumped solid-state (d.p.s.s.) laser effectively began an escalation in the availability of power to the holographer in the 1990s. A laser of this type works by the most simplistic frequency-doubling mechanism from the exceedingly powerful Nd:YAG crystal, whose naturally efficient output line is in the infra-red at 1064 nm.

As shown in Figure 3.11, a powerful red (semiconductor) diode laser is responsible for pumping the lasing crystal itself. This laser beam is focussed into the lasing crystal through a wavelength-selective reflective coating which is highly reflective of 1064 nm light generated in the crystal, so that little or none of the primary stimulated laser light is allowed to escape back toward the pump laser. As a result of the ongoing pumping, rays of infra-red light cycle inside the laser cavity in the classic lasing action. Included inside the laser cavity, close to the laser crystal, is a non-linear frequency-doubling crystal whose "output-coupling surface" is coated with a dielectric layer which selectively reflects light at 1064 nm, but transmits the frequency-doubled 532 nm green component of the energy. Infra-red light thus remains within the laser cavity, returning to the pumped lasing crystal where ongoing amplification continues.

Further focussing optics collimate the green output beam to ensure that a very small (<< 1 mm) TEM00 beam of very low divergence is emitted. Because of the stability of the optical cavity, immense values of coherence length are available from d.p.s.s. lasers.

Swedish laser engineers Cobolt [8] have produced a whole series of lasers of the d.p.s.s. type which have proved useful for hologram mastering. They are compact, reliable, have a long life expectancy and a low-current, low-voltage power supply in a small, separate unit. They need a small, air-cooled heat sink. For the lower power lasers, even without cooling fans, this does not run hot to create air currents which disturb the holography stability requirement.

This technology offers a wide range of visible laser colours. Blue wavelengths close to the time-honoured lines formerly available from the argon gas lasers (458 nm, 476 nm

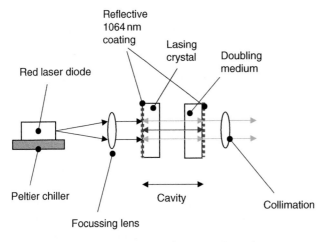

Figure 3.11 Schematic operation of a 532 nm d.p.s.s. laser.

and 488 nm) are now included in the range of wavelengths offered by this form of laser, at nearby lines of 457 nm, 473 nm and 491 nm. These all have a significant coherence length for holography. The interior of the compact laser housing is literally a master-piece of delicate miniature optical engineering – truly a sight for sore eyes!

At the expiration of a 35 mW helium–neon laser in 2011 at our laboratory at 3DOS in Halstead, we noted the absence of the line equivalent to helium–neon from the portfolio of d.p.s.s. lasers. The nearest 659 nm line available in d.p.s.s. lasers at that time was a deep red, which has a photopic disadvantage for visual holography. The helium–neon 633 nm line has had a long and very successful history in holography as it is such an attractive colour.

Upon making enquiries to Lambda Photometrics [9] for a replacement from LASOS [10], we found, to our horror, that the traditional HeNe laser had been discontinued.

Lambda organised a meeting at Harpenden with LASOS and I was able to put the case for the development of a d.p.s.s. laser with an HeNe "lookalike" wavelength and improved coherence. LASOS listened intently, and a few weeks after the meeting affirmed their intention to create the RLK4075T – a red d.p.s.s. laser at 639 nm wavelength.

Red lasers following this prototype are now available at 200 mW and are the perfect replacement for the much-loved helium–neon lasers which were formerly the bedrock of so many holography studios.

An additional advantage of the Cobolt and LASOS lasers is that they have the ability to connect to remote laptop software which is capable of adjusting the power and current controls. This makes for a very convenient way to adjust output power of the laser, and the safety implications are clear when using the laser to set up an optical configuration on the holography table at low power. Both of the manufacturers, LASOS and Cobolt, have recently demonstrated their ability to tune these d.p.s.s. lasers online through the "Team-Viewer" software with complete success. After years of experience of delicate manual "walking-in" procedures for gas lasers, the sight of the laser beam being adjusted live from another country can only be described as surreal!

3.14   Fibre Lasers – A Personal Lament!

Perhaps the most frustrating "non-event" of my own technical career, which has been spent entirely in the pursuit of technology associated with light, is my own failure to recognise the obvious potential of optical fibres in the visible laser field.

I was involved in a successful period of intensive research into the vapour deposition of low-loss graded index fibres at Standard Telecommunication Laboratories [11], but, after becoming involved with holography, despite my knowledge that the pioneers of the ruby laser faced such a difficult task to find ways of introducing sufficient pump light into the cavity, I now find it truly staggering that I did not recognise the fact that the optical fibres of the type I had recently made on the frontier of that technology presented an ideal medium for a very narrow laser cavity of considerable length.

This was despite the fact that in the earliest days of my 14 years at Applied Holographics, I actually discussed the principles of fibre-based light gyroscopes with my late cousin Peter Howard, who was a leading engineer with British Aerospace and a brilliant physicist. The glass fibre presents a high numerical aperture for the admission of pump light,

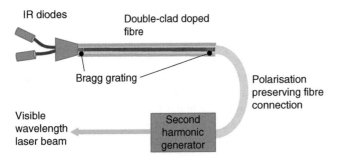

Figure 3.12 Schematic for the visible fibre laser.

and the manufacturing process I was using personally was eminently suitable for selective doping of the silica with metals such as Ytterbium etc.

Nowadays, the doping of optical fibres with chemicals at low concentrations is able to provide a number of frequencies which are suitable for frequency doubling. Clearly, the ability to produce lengths of fibre allows an incomparable opportunity for pumping. Unfortunately, really long laser cavities introduce an expanding possibility of multimodes, so currently the usable length of fibre is limited, but it is, of course, possible to coil the fibre to enable a compact enclosure.

The central core of the fibre is terminated at each end with Bragg gratings of the appropriate fringe spacing, which effectively bound the cavity, as shown in Figure 3.12. These gratings selectively reflect the desired wavelength component at the exclusion of the pump frequency. Additional pump light can travel in the outer core but the principal inner lasing cavity continues to oscillate at the selected principal wavelength. The fibre can feed a second harmonic generator in much the same way as occurs in the d.p.s.s. laser previously described.

Fibre lasers are a rapidly developing technology which promises to deliver high power, long coherence and high stability in future. They have the great advantage that the heat-generating pump sources are remote from the SHG, which is effectively the "laser head". Furthermore, the ability to dope the fibre with a range of rare earths, etc. promises to provide a useful range of frequencies for visible lasers.

3.15   Eye Safety II

Where possible in holography, we need to create machines and systems where all laser beams are enclosed inside a sealed casing to avoid the operators being exposed to laser light, especially in view of the fact that a full-colour system makes it difficult or impossible to rely upon spectral filter safety goggles for protection. However, the operation of conventional master making is a visual process and we need to use advanced solutions with carefully selected safety glasses to allow manual operations to be carried out with the required precision whilst retaining safe practice for the technician. The nonchalance towards eye safety which senior holographers often tend to show as a result of their early experience (of being limited to working with a few milliwatts of laser power) *is not justified in modern laboratories which benefit from the new laser developments*.

Whereas multi-colour holography has apparently made laser safety issues even more complex, we are fortunately moving into an era where software-controlled cameras can be used to assess the set-up quality of optical tables.

3.16   The Efficiency Revolution in Laser Technology

One of the great problems with the use of the gas lasers HeNe, argon, krypton and HeCd in past years, especially for the argon and krypton lasers, was the difficulty in providing cooling facilities and power supply. The infrastructure of a lab with three-phase power supply and plumbing to cool an ion laser became a major headache in the past; this is rapidly evaporating in this decade with the new generation of d.p.s.s. lasers.

We have experienced in the past significant difficulties in constructing and maintaining cooling systems for gas lasers. At Applied Holographics, a typical installation in the 1980s related to a system which was designed to produce 6″ square photoresist masters. This was a krypton laser working in the violet line at 413 nm and capable of 0.9 W. A three-phase power supply in the form of a 30 mm diameter cable ran hot to the transformer and onwards to the laser head. This level of input power was providing "just" 0.9 watts of laser power, albeit at that time a breakthrough in the exposure of photoresist, for which Nigel Abraham deserves great credit; a development which allowed AH to expose 8″ × 10″ resist plates with realistic exposure times. Figure 3.13 is an example of an excellent embossed hologram [12] which was thus facilitated, shot as a single gang on an 8″ × 10″ resist plate.

Unfortunately, however, the remainder of that three-phase input power was necessarily absorbed into cooling water, and the vital requirement for temperature control of the laser cavity and the etalon assembly meant that accurately controlled water flow was required. Basically, a two-loop process was used, where a precision heat exchanger with accurate flow control was linked into an external chiller unit; this needed preferably to perform approximately equally in summer and winter, despite the significant extremes of ambient temperature to which it was exposed.

3.17   Laser Coherence

The phenomenon of "coherence" must be divided into two categories. *Spatial coherence* is defined, in effect, by the size of the source; *temporal coherence* refers to the extent to which the waves of light are absolutely monochromatic and precisely in phase with one another with respect to time.

Of course, with laser light, it appears that we can almost take this for granted since we have monochromatic light – but for holography, we ask not only for "single longitudinal mode" (SLM) but for *extremely* narrow bandwidths; wave bandwidths of ever-narrower distribution now offer us modern lasers with "coherence lengths" of many metres.

For the holographer, this means that with a properly set optical arrangement, we can easily make holograms of large objects. It also means that a certain level of ill-discipline in setting the optical table will be tolerated; however, we should avoid such laxity if we

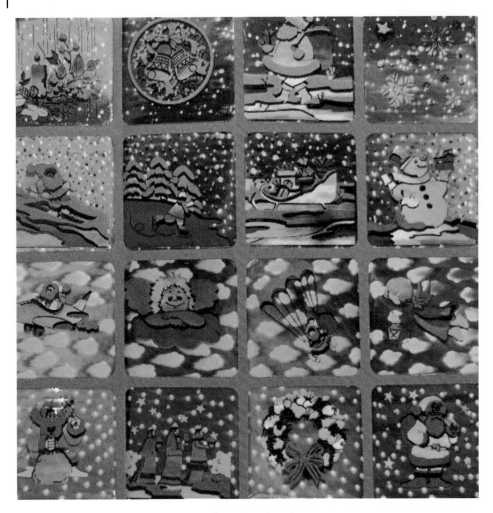

Figure 3.13 Image-gang hologram exposed at 413 nm in resist.

are to make sure that the system always works perfectly when the tolerances of conditions and equipment performance are stretched to the limit by other factors – and they often are!

Realistically, the TEM00 classification is the first requirement in the selection of a laser for conventional holography. This term refers to the profile of the laser beam and is a function of the configuration of the optical cavity itself. A predominantly circular output beam from the laser is the characteristic of the TEM00 condition.

But the definitions of bandwidth, which controls effective coherence, and the difficulties in providing perfectly stable cooling facilities, are difficult to assess for the holography application and there is no more certain means to ensure successful application of ever-changing laser technology than to carry out practical testing prior to the acquisition of a laser. We detail such tests in Chapter 8.

Notes

1 Emilio Segrè Visual Archives: American Institute of Physics.
2 Hecht, J. (2010) *Beam: The Race to Make the Laser*. Oxford University Press (ISBN: 978-0199738717).
3 LaserMet www.lasermet.com
4 Coherent UK Ltd www.coherent.com
5 Laser Lines Ltd, Beaumont Close, Banbury OX16 1TH.
6 Melles Griot www.mellesgriot.com
7 Kimmon Koha www.kimmon.com
8 Cobolt www.cobolt.se
9 Lambda Photometrics www.lambdaphto.co.uk
10 Lasos www.lasos.com
11 Standard Telecommunication Laboratories was the UK Research Laboratories for the Standard Telephones and Cables Company (STC). It is now recognised as the birthplace of optical fibre communications, for it is here that Sir Charles K. Kao, George Hockham and others pioneered the use of single-mode optical fibre made from low-loss glass. In 2009, Kao was awarded the Nobel Prize in Physics for pioneering optical fibre communication.
12 Christmas hologram image by kind permission of Applied Holographics/Opsec www.opsecsecurity.com

4

Digital Image Holograms

4.1 Why is There Such Desire to Introduce Digital Imaging into Holography?

With the radical developments in IT and communications of recent years, computer graphics has moved into the domain of the typical amateur. The photographic and printing industries have come under intense pressure from technological developments to advance towards digital imaging methods as far as is possible.

The remarkably lengthy search we may now experience in shopping for a roll of colour negative film reminds us just how quickly digital imaging equipment has taken the place of the traditional film camera in the consumer market, and many homes have inkjet printers capable of producing first class prints direct from a digital camera or graphics software.

This same driving force impacts upon holography, but in this case the basic definition of the original concept of the process (*the entire message*) with the ramification that "every part of the recording contains all of the image" is essentially at odds with the familiar principle of dividing the surface into a matrix of discrete imaging zones which individually contribute to the total aggregated graphic image.

Furthermore, the very essence of the original *holography* concept was that the reconstructed image was not necessarily in the plane of the recording material, and thus any individual point on the surface of the film does not coincide with a single, consistent property of density or colour. However, some of the digital systems we shall discuss do, indeed, deal with this apparent anomaly.

Even in the case where the basic artwork itself is of a digital format (for example, in the photographic recording process of the individual frames of a holographic stereogram sequence captured with a high-resolution digital camera), the conventional ("analogue") recording process involved in making a master hologram by means of classical holography still has much to offer. The first generation master (H1) is, in principle, able to record a seamless sequence of incredibly high-resolution component images of a real scene. In this case, we are making an analogue recording of digital information and arguably taking the opportunity to get "the best of both worlds".

But clearly, the "artisan" nature and reputation of holography implies, and to some extent enjoys, high origination cost and long timescales typically associated with traditional trades involving such craftsmanship. There are parallels with banknote printing,

The Hologram: Principles and Techniques, First Edition. Martin J. Richardson and John D. Wiltshire.
© 2018 John Wiley and Sons Ltd. Published 2018 by John Wiley & Sons Ltd.
Companion website: www.wiley.com/go/richardson/holograms

where the highly skilled "engraver" trade arguably survived far longer than was the case in other areas of printing. It is no surprise that the link with that trade was pursued within the De la Rue Holographics group, who have produced over many years relatively conventional "2D/3D" holograms whilst offering the very highest image quality, due to their remarkably rigorous master origination process.

However, the era of digital holography appears now to welcome every graphic designer to the fray of hologram creation! Coupled with this explosion of design talent waiting to contribute to holography, there comes a commercial need to shortcut the hologram origination process to achieve a rapid turnover of new artwork converted into holographic form, and a desire to eliminate the delays and blind alleys which inevitably present themselves in the conventional complex multi-stage hologram mastering processes; a "one-step" process sounds ever more attractive.

Thus, for both practical and philosophical reasons, both the security and display limbs of the hologram industry have advanced towards an ability to produce sophisticated new images in a modern form involving digital graphics and process automation.

In the case of artistic display holography particularly, my co-author is intensely aware of the vital need for holography to retain its reputation as a futuristic technology; this implies the exploration of modern trends for interactivity, compatibility with virtual reality (VR) systems and, eventually, full rapport with artificial intelligence (AI) concepts.

The security print industry has taken the concept of optically variable devices (OVDs) very seriously indeed. In essence, the iridescent colours and animated effects of these diffractive devices cannot be simulated by conventional printing processes and they automatically set a new challenge to the counterfeiter. The more specific term DOVID (diffractive optically variable image device) differentiates the diffractive methods of producing such effects and a further category of ISIS (interference security image structure) exists which refers to thin-layer interference effects, ranging from simple, single-layer coatings up to complex, multi-layer stacks (and surely therefore can be considered to include volume reflection holograms).

Rudolf van Renesse, in his book *Optical Document Security* [1] provides great detail of the various security systems which rely upon diffractive systems and shows useful samples of the techniques. The book also explains many wider aspects of security print, biometrics, etc.

4.2   The Kinegram

Landis and Gyr invented the Kinegram in the 1980s and this could realistically lay claim to being the first commercial application of digital holography. However, this was certainly not presented at that time as a method to reduce costs; rather, it was always (and still is) seen as a most sophisticated method of protecting against counterfeiting aimed principally at governmental and banking applications, and this was reflected in the price of origination.

The detailed technical explanation for this famous and highly reputable process has been well protected from the public gaze and remains shrouded in mystery to this day; it is often a subject for detailed speculation and discussion amongst experienced holographers over coffee! How did Landis and Gyr devise a digital system based upon vector graphics at such an early stage in the development of the security holography industry?

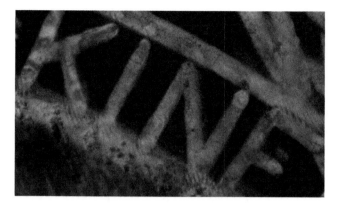

Figure 4.1 Photomicrograph of "Kinebar" hologram.

In essence, the Kinegram process is a high-resolution vector graphics system with line spacing in excess of 1000 lines per millimetre. Figure 4.1 shows a photomicrograph of such an image in the famous application where the technology was used to decorate and authenticate solid gold ingots. Since that time the basic "Kinegram" technique has been used in many of the world's most prominent security applications including banknotes and passport overlays.

Landis and Gyr eventually released the ownership of the technology to OVD Kinegram [2] of Zug, Switzerland, under the control of the German foil and print experts, Leonard Kurz. Since its inception, the company and its technology have achieved an outstanding reputation in security circles. The technique is held in high esteem and has remained shielded from any attempt to claim parallel methods. The Kinegram image comprises minute zones of diffractive surface-relief gratings. Each adjacent zone is configured with a very slight incremental variation in the angular orientation of its relief fringe structure, with the effect that the viewer's eye is stimulated by different areas of the foil surface as they move with respect to the lighting. Thus, as the name suggests, the viewer perceives motion in the image graphics as the foil, the eye or the light source moves. To a great extent, this morphing or animation effect has been selected as a high-security feature, arguably *in preference and alternative* to the three-dimensional capability of conventional holography.

The Kinegram technique offers a highly sophisticated appearance combined with resolution characteristics capable of producing microtext in vector form. Diffractive microtext is a favoured feature in the security industry. Such minute detail is extremely difficult to achieve in conventional holography due to laser speckle and general difficulties of preparing and efficiently imaging such detailed artwork with lasers.

The intermediate product of the embossed holography origination systems, including conventional holography and the digital techniques such as the Kinegram, is a metal shim that is capable of storing a surface-relief recording of predominantly linear diffractive microstructure, which can be impressed during the embossing process into thermoplastic foil. Such foils are frequently coated with purpose-designed lacquers which facilitate the transfer of relief structures between generations. These lacquers are the key to the mass production door, because it is also possible to coat similar lacquers onto rigid carriers.

This is the technique that is used in the recombination, or the step and repeat processes, used to produce matrix gangs of small images for use as embossing shims, for mass replication of small images, which may be either conventional holograms or digital images, or a combination of both.

The techniques used by experts in the recombination field, such as Wavefront Technologies Inc. [3], to produce arrays of holograms in their routine work use a similar method to add small component gratings to act as registration marks. Such groups working in the "recombination" of individual images have produced specialist polymer lacquers to receive the surface-relief patterns in high resolution at high speed, and these materials can be coated upon plastic, glass and film layers. Lacquers which are cured after pressing with ultra-violet exposure have proved to have excellent properties of fringe replication, to the point of actually *improving* the original relief profile! This possibly indicates a general route by which microscopic zones of diffractive microstructure could be aggregated into a vector matrix of the type we see in the Kinegram.

4.3 E-beam Lithographic Gratings

The vector-based Kinegram method described above differs substantially from the e-beam lithographic and other digital methods, wherein the surface is divided geometrically into simple pixels or into linear "tracks", which are grid-like surface sub-divisions. The Commonwealth Scientific and Industrial Research Organisation (CSIRO) in Australia has been the field leader in the use of electron-beam lithographic technology. Robert Lee [4] was the inventor of the techniques featured in the various generations of CSIRO's work. The original image system was known as "Catpix" and this method was transferred to the Reserve Bank of Australia and produced the well-known Captain Cook hologram in a transparent window on one of the early plastic banknotes from RBA in 1988.

However, many scientists in the former Eastern Bloc have also created systems for e-beam lithography with the ability to write high-resolution diffractive structures into resist materials which are sensitive to e-beam exposure rather than the blue-light-sensitive resists used for conventional holographic recording.

After the sale of the original Catpix specification to Reserve Bank of Australia, CSIRO's new advanced method of e-beam lithography was called the *Pixelgram*. Each individual zone contains microstructure in the form of individual linear surface-relief fringes *written directly into photoresist by electron beam lithography.*

These zones were, in the first iteration, an orderly array of rectangular pixels. Within each pixel, a guided beam of electrons is able to draw linear fringe structures which are not necessarily uniform, and, in fact, which may be either predominantly linear or predominantly curved, with the effect that such pixels act either as angled plane mirrors or, in the case of the pixel containing a curved grating, producing a divergent cone of reflected light. This dichotomy of structure gives rise to a unique headline feature for the CSIRO product – namely, that any image can be made to switch between positive and negative modes by a small degree of tilting action.

Later, the *Exelgram* was introduced as another development of the process. Elongated tracks replaced the original pixelation but again the fringes written within these individual zones were variable. The Exelgram has the advantage over the original configuration that the problematic diffraction and scatter at the pixel edge is much reduced by removing such a large proportion of the boundaries within the image array.

One of the advantages of e-beam work is its very high resolution in comparison with the techniques where visible light is used to address the recording material; the ability to write individual diffraction fringes. Unlike the holographic process, there is no deleterious laser speckle to detract from the image quality; this is especially advantageous for "microtext" features.

In fact, the resolution of the etching process nowadays enables fringes to be written with characteristics associated with classical wave profiles – cosine, triangular, saw tooth, square wave, etc. This immediately raises the issue of blazed gratings, whose asymmetrical profile can have the effect of increasing the proportion of light diffracted into the first order.

The difficulty for the e-beam protagonists is that the appearance of the Exelgram *to the layperson* resembles the Kinegram and dot-matrix technology. To the experienced technical eye, the resolution, sharpness, colour quality and other design features of high-end devices are recognisable, and certainly at the forensic level these features are simple to detect, but as regards counterfeiters' attempts to fool the casual viewer (whose banknote fidelity is their own responsibility), the dot-matrix machinery which is now widespread across the world, and in some cases belongs to unscrupulous owners, is capable of producing lookalike images of very similar appearance to the layperson.

Actually, the ability of the layperson to analyse an embossed hologram is shockingly poor. For this reason, the appearance of a silvered diffractive image, whether originated by the elaborate and sophisticated e-beam writing system or by the low-cost dot-matrix system described below, is, to some extent, indistinguishable to the typical viewer. In fact, two of the first "counterfeit" issues that I was personally asked to comment upon were simply foil patches with no diffractive qualities at all. Our industry has a significant quantity of work to do in terms of educating the general public!

Ironically, the classical holography method of "2D/3D" where graphics images are separated into several discrete layer components, which are represented at differing depths within the hologram image, tends to differ in appearance from the digital methods, which may merge in the perception of the interested viewer.

4.4   Grading Security Features

In security print there is a general agreement to summarise features as:

- Level 1 – front line visible features for the awareness of the layperson.
- Level 2 – verifiable features including covert information.
- Level 3 – features offering forensic confirmation.

Holographic features naturally tend to inherit these definitions when included in security documents, so that the Level 1 category is literally the inclusion of a hologram as a visible phenomenon.

However, over the years, counterfeit banknotes have typically included blank hot-foil patches, foil patches with debossed or printed features and, of late, actual counterfeit versions of a hologram. There was a particularly important "landmark" counterfeit hologram of the €200 note, which was of good quality and thus significantly damaged the reputation of embossed holography.

In my experience, the general public takes a surprisingly casual view of the existence of counterfeiting in the banknote sector. In my work in the past, I heard tales of people taking fake notes into a High Street bank to ask for them to be exchanged! The reality is somewhat different – one is effectively in default of the law by attempting to use such a note. The situation can be particularly complicated for holidaymakers in Europe, who have actually been arrested for possessing counterfeits passed to them in a batch of notes from a tour company. So it would be to everyone's advantage to try to expand their knowledge of each security device, including the hologram. But such information must come from the holography industry. There are other areas of security where not only money is at stake – holography is routinely nowadays used to protect medicines and engineering components for automotive and aeroplane brakes...education, education, education!

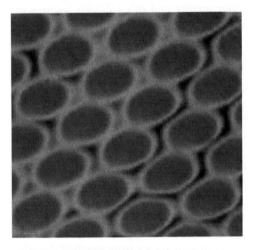

Figure 4.2 Compact dot matrix.

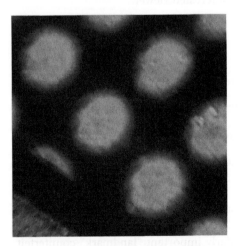

Figure 4.3 Spaced dot matrix.

It is possible to include covert features in holograms which can be verified with specialist devices. A good example would be in the holograms for tickets and clothing labels made by De La Rue Holographics [5] for the German World Cup in 2002. Interrogation of certain zones of the surface with a small pointer laser reveals a clear projected image – "FIFA" – which can be captured on a nearby screen.

At the forensic level, it would be extremely difficult or impossible to produce a counterfeit which would stand up to all of the tests that a laboratory could apply, without clear differentiation from the original.

In Figures 4.2 and 4.3 we see that simple optical microscopy, for example, will normally reveal detail which is specific not only to the equipment used, but to its condition of operation during a particular origination process. This is illustrated by the two figures which show configurations from two competitive "dot-matrix" systems.

Ironically, such minor variable qualities in the original imaging process seem to mean that, at the forensic scale, a level of identification of the source of a hologram exists which offers precise origination location and even date information, since beam drift tends to occur, producing characteristic imperfections in the image dots, as shown in Figure 4.6, which realistically is absent from the more perfectly crafted e-beam work.

Opsec Security Ltd [6] has developed systems for specific encoding of the individual dot.

For all its advantageous qualities, it appears, with hindsight, that dot-matrix origination technology has ironically, to some extent, undermined the security value of embossed holography. The overwhelming problem is that despite the efforts of prominent manufacturers such as Applied Holographics (Opsec) to avoid the situation, equipment from other manufacturers, sometimes in the form of "second-hand" ownership, has found its way to parts of the world where it has been directed towards the compromise of legitimate security holograms. This means that the relatively inexpensive dot-matrix equipment is very attractive to the unscrupulous supplier of embossed holography.

It is certainly the case that a highly skilled and experienced dot-matrix designer/operator can utilise the design software to simulate many of the visual effects which the high-end techniques utilise.

4.5   The Common "Dot-matrix" Technique

Applied Holographics plc. was one of the pioneers of "dot-matrix" holography. In the mid-80s, Craig Newswanger, one of the great inventors in the realm of scientists who have graced holography and technical guru for AH Inc. (later AH Corp.), worked with Nigel Abraham, representing AH plc., to introduce to the Braxted Park, Essex, laboratory components for a rudimentary system for the production of digital master holograms in photoresist. These masters comprised an array of minute, individually exposed pixels, whose surface-relief gratings had the property of diffracting incident light through a range of angles of rotation, as shown in Chapter 1 (in Figure 1.12).

These plane gratings were essentially very simple uni-directional linear surface microstructures which were capable of high diffraction efficiency – principally as a result of the simplicity of the grating structure of each discrete pixel. The superficial graphic design of the hologram was therefore divided into various miniature zones and line patterns which differed in the rotational angle through which they reflected incident light.

Thus, a viewer looking at this type of hologram from a certain position will see a mirror-like reflection from those pixels which contain a grating whose fringe orientation directs reflected light towards their eye. Pixels which direct light elsewhere appear dark. Of course, the dispersive grating will introduce colour effects as the viewer or the source of illumination moves, and, most importantly, as the hologram is rotated through a range of angles, different zones will illuminate and change colour.

The astute designer will be able to create motion, polarity reversal, emphasise text or logos against their backdrop etc., all of which will appear to the viewer to be, to some extent, interactive. If the viewer moves their viewing position or rotates or tilts the surface bearing the hologram, there might be an apparent flowing or filling motion.

The basic mechanics and optical system of dot-matrix machines are relatively simple in principle, as shown in the schematic in Figure 4.4.

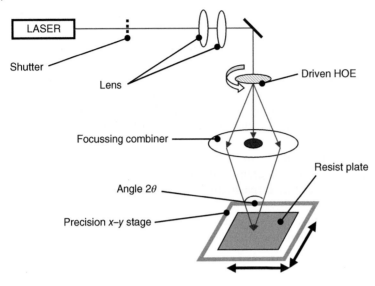

Figure 4.4 Dot-matrix system.

The laser beam is blocked by a shutter system which needs to operate quickly and repetitively. It is fascinating to recall that the original mechanical "bi-stable" shutter selected by Andrew Rowe and myself for original research at Applied Holographics was specified as being capable of "millions of cycles" but after its acquisition, we rapidly realised that it would shortly be cycling millions of times every day! Later, the use of opto-acoustic shutters provided a speed advantage as well as long-term durability. These shutters are very difficult to adjust for 100% beam attenuation. In some cases, dot-matrix systems have, therefore, utilised an over-riding mechanical shutter to block the intense laser beam during lengthy periods where the photoresist recording plate is in position when the exposure system is at rest.

The holographic beam divider ("driven HOE") is a linear grating which simply divides the on-axis incident laser beam into two first order parts. It is rotated between expo-sures to provide a range of grating directions. The zero order beam is absorbed at, or before, the combining element, which may be a conventional lens or a holographic component. The focal point of this converging optical component coincides with the photoresist surface, and in conjunction with the meniscus lens focusing system and the beam divider, provides a single focussed dot comprising two angular components in the precise plane of the photoresist.

The first lens or lenses provide a focussing facility. These are meniscus lenses of very long focal length which may be used to adjust the dot size. The size of the image dot, and thus the resolution capability of the image, is reduced in accordance with the incremental tracking distance of the tracking system that holds the recording plate in order to achieve maximal surface coverage with dots containing diffractive grating (Figure 4.5). During the focussing process, the dot is examined microscopically and the effect of rotation can be assessed with regard to the shape of the dot produced by the two coincident coherent beam components.

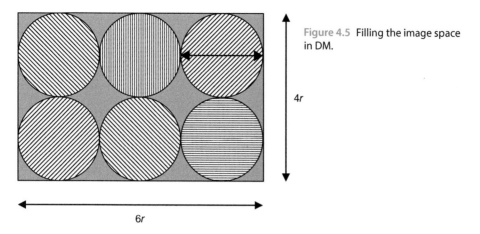

Figure 4.5 Filling the image space in DM.

4r

6r

The setting up process comprises minute adjustments of the registration of the two beams such that the spots of light from each beam are precisely overlaid and that their size coincides with the tracking increment.

For circular dots, as a guideline (in practical terms the dots are ellipsoid) the coverage is:

Area of a diffractive dot $= \pi r^2$

Rectangular area of photoresist $= (2r)^2$

$4r^2/3.142\,r^2 = 4/3.142 = 0.7855 -$ Say 80%

Some dot-matrix systems are able to produce a "square dot" with which to fill each pixel exactly in order to ensure greater surface coverage. Of course, the additional 20% surface coverage is balanced against any loss of mean diffraction efficiency of the gratings forming the dots – the manipulation of the obvious circular configuration is inevitably responsible for losses. As mentioned, Opsec has manipulated the dot shape to the extent of encoding specific security information in the individual dots.

In fact, the first dot-matrix holograms that we made at Applied Holographics had a resolution of only 70 dpi. However, the need to accommodate very small text in security holograms was the driving force for continued reduction of the pixel size. A natural immediate target was to simulate security printing, which has typically contained "microprint" of the order of 0.25 mm in height, so the necessary five bars of the letter "E" are thus of minimal dimension 50 μ.

Figure 4.6 shows a typical non-vector microtext letter of the type common in dot-matrix work. This differs clearly from the Kinegram text previously shown but is clear and bright to offer ease of legibility to the viewer under various lighting conditions, since the gratings are locked in position in the surface and do not suffer from the blurring effects which might afflict analogue holography.

If we use a blue helium–cadmium laser with a wavelength of 442 nm, then the Bragg grating equation can define the fringe spacing which exists in the linear grating:

$\lambda = 2d \sin \theta$

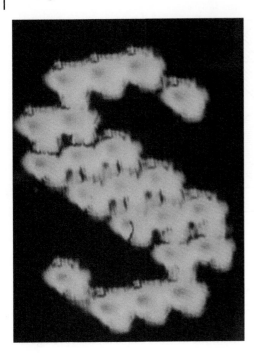

Figure 4.6 Microtext.

Then $\lambda = 442$ nm; if half the angle between the incident beams, θ, is, say, 20°, then the fringe spacing will be:

$$442 = 2d \sin 20°$$
$$d = 646 \text{ nm}$$

Gratings of the frequency dictated by this ~0.65 µ fringe spacing in this type of "hologram" are distributed with a range of axial orientations through a full ninety degrees. The complex and multi-disciplinary process of conventional hologram design is effectively eliminated here, and a new breed of skilled designer, working closely within the familiar environment of computer graphics, has been able to produce stylish OVDs which provide iridescent colours and animation effects whilst filling the whole surface of the embossed foil in such a way that the common criticism of conventional holograms, as to their restricted viewing angle, is completely eliminated.

It is certainly the case that a dot-matrix hologram, left lying on a table, will almost always display an interesting flash of colour to invite the interest of the casual viewer from any position in the room, which is not necessarily the case with the conventional rainbow hologram. There is a fascinating anomaly, shown in Figure 4.7, whereby the individual dots become involved in "cross-talk", which leads to further loss of diffraction efficiency resulting from the inability of circular dots to fully utilise the whole of the available foil surface. Under certain conditions of exposure, there is interference between adjoining image dots. Of course, as mentioned previously, we are in the realms here of fortuitous forensic features which could literally never be counterfeited – a claim which has often been made in holography, but which, in this case, seems justified, although "dot matrix" has often been considered the poor relation of the Kinegram and e-beam techniques.

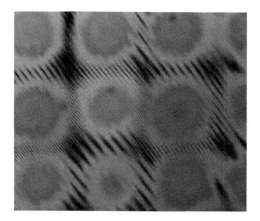

Figure 4.7 Cross-talk between dots.

Certain other dot-matrix systems, which can also work well, utilise a more powerful laser, but simply permit the laser light from a converging lens to focus at a single point on the photoresist recording plate whilst a rotating circular mask plate provided with two apertures at the appropriate places delivers the laser light to the resist for each exposure. This has the advantage that a range of grating frequencies can easily be offered, but the disadvantage that the lower level of energy at the resist extends the optimal exposure time and, accordingly, the stability and settling time requirements for each dot component, thus increasing the time taken to finish each resist master.

Although the photoresist materials universally used to originate embossed holograms are relatively slow in photographic terms, the individual exposure dot is very small indeed, as the focal point of two very carefully aligned and focussed laser beams. Typically, a helium–cadmium or d.p.s.s. laser of low energy at 442 nm is used. Shipley photoresist was designed originally with peak sensitivity suited to the mercury arc spectrum line at 436 nm, so the 442 nm line of HeCd is ideally suited to the material; more actinic, in fact, than the argon 458 nm line which has often been used for more conventional embossed holography work.

The consequence of a spot of light of ~50 microns diameter comprising two overlapping beams efficiently delivered from a low-power (say, 30 mW) laser with minimal losses, is that the required exposure time to the individual dot is just a few milliseconds, and, accordingly, there is a need for very little settling time before the shutter opens, after the photoresist plate arrives at its position for each sequential pixel exposure. In consequence, the exposure of a typical 25 mm square image unit at > 300 dpi is completed in no more than a few hours, comparing favourably with conventional holography.

After processing the photoresist plate, it is coated either electrostatically or chemically with silver metal. This is a remarkable experience to witness, as the surface-relief image recorded in the deep red surface of the iron-oxide-backed photoresist plate is suddenly converted into a highly reflective silver diffraction grating. A nickel shim is then grown by depositing nickel on the thin, metallised surface from a nickel sulphamate plating bath to produce a nickel shim of the desired thickness.

This "mother shim", or copies from it, can be used in mechanical step and repeat processes to produce a precisely registered multiple image shim to be used for mass production purposes. A single image area, typically 25 mm square, is thus used to create an image gang of, say 150 mm square. This process, as executed by Opsec etc., is sufficiently precise that it is often impossible to see with the naked eye the repeat boundaries in the compound image.

In the second generation of size expansion, the 150 mm multiplex image tiles may be formed into an array to run on a wide-web embossing machine in excess of one metre web width.

4.6   Case History: Pepsi Cola

There was a fascinating example of commercial pressure influencing technology objectives at Applied Holographics in 1994. Applied had achieved a reliable day-to-day routine for the production of dot-matrix holograms via a "step and repeat" process from a small unit to build a 150 mm shim, as previously described. Pepsi demanded a label for the family-sized bottle of its product, which was required to cover the full label with a *single image.*

The dot-matrix design was a continuous image featuring animation effects reminiscent of a firework display. Pepsi wanted the "firework" spray animation to be continuous across the whole label.

There was no option but to run the 6″ square, but technical staff pointed out that it would take 4.5 days. The local electricity supply company seemed incapable of giving Braxted Park continuous power for more than a few days without cuts. This was before the general introduction of "battery back-up" laptop computers, so we had already organised a power buffer support system on the desktop control system which could keep the computer going through an extended power cut; but it couldn't supply the helium–cadmium mains laser itself. In fact, the machinery delivered its pixel exposures at an enormous rate and so any loss of laser power would result in the loss of a significant number of pixels even if software had been developed to stop the system rapidly in the event of a loss of laser power. (Of course, nowadays it would be relatively simple to organise a battery back-up for a d.p.s.s. laser.)

Naturally, for a high-profile commercial piece of work, everyone in the company was keenly aware and interested in the hologram running in the dot-matrix lab and it ran for more than three days before the first abortive power cut. The pressure on the timescale mounted. After the re-start we were fortunate enough to get a continuous supply for a week, which was sufficiently long to complete the Pepsi hologram, but at completion of exposure, the pressure on the chemistry department to process such a unique and almost irreplaceable plate was severe. Large resists are much more difficult to process than smaller plates for reasons described later.

The production embossed hologram labels on the Pepsi bottle were received successfully and won awards in the packaging industry for Applied in 1995.

4.7   Other Direct Methods of Producing Digital Holograms

There are a number of hybrid methods whose advantage over conventional dot-matrix work is that the individual surface pixels or lines are encoded with far more complex data, with the effect that they portray high-quality, colour, three-dimensional animated images with high resolution and high brightness with a really pleasing aesthetic appearance; all with the overwhelming advantage of a "one-step" origination process.

It is useful to appreciate that the method which is at the root of these digital 3D processes involves the anticipation of the full range of component views of a 3D subject from various viewing positions which the viewer may assume in relation to a single viewing point in a plane (the film) within the image space itself.

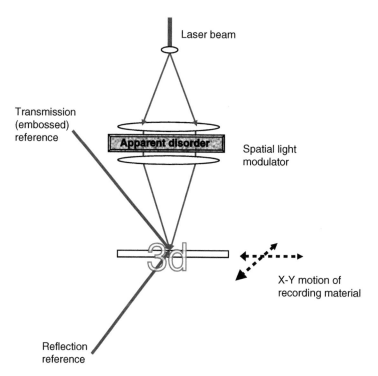

Laser beam

Transmission
(embossed)
reference

Apparent disorder

Spatial light
modulator

X-Y motion of
recording material

Reflection
reference

Figure 4.8 "One-step" digital origination system.

So, for every point on the surface, a "ray-trace" calculation occurs as to what the viewer's eye, in its displaced position on the base of a pyramid of which this particular pixel is the apex, *should* see in that line of sight when contemplating a three-dimensional scene straddling the recording plate.

Thus, if every pixel presents a realistic panorama of views from its position in the matrix, the binocular vision of the viewer will interact in such a way as to allow the perception of stereographic vision. In volume reflection holograms, there is also an opportunity for both parallax and animation effects in the vertical axis.

Thus, the imaging device, which is a spatial light modulator whose output is focussed through an imaging system towards the pixel on the surface of the film, separately encodes every individual surface point with the resulting computer-generated information in three colours. In transmission holograms, with no parallax in the vertical, the display is banded, but in reflection applications, the graphic display image is two-dimensional. One can imagine that the image on the spatial light modulator during the recording of each pixel in the grid matrix, which represents the view emanating from the individual image component point, makes for an apparently illogical graphical scene when viewed in isolation. This principle is shown schematically in Figure 4.8.

The focussed light from the SLM is met at the film surface by a focussed coherent reference beam, recording a pixel grating capable of playing a part as one component of a matrix of square pixels representing a three-dimensional scene. Subject matter "in front" of the film will thus "shadow" the effect of subject matter "behind" the film.

4.8   Simian – The Ken Haines Approach to Digital Holograms

Ken Haines [7] has a pedigree in holography which began at the time of Leith and Upatnieks' vital "off-axis holography" invention. With his rich appreciation of the phenomenon of three-dimensional imaging, his approach to early digital holography was to divide the surface of the recording into a matrix of pixels, each of which has the ability to represent a position within the volume of a three-dimensional image.

Working with his daughter Debbie, who is skilled in programming, a complex system was devised which offered perhaps the first commercial "one-step" method to create three-dimensional holograms. Like the simple dot-matrix systems described above, this technique allowed the graphic design to be entered into the system in such a way as to eliminate the need for any significant technical expertise in the holography laboratory, with the exception of chemical processing of the automatically exposed photoresist plate (as described later in Chapter 6, this is, in itself, not insignificant in terms of expertise!).

In this system, as outlined above, the individual image pixel on the surface of the photoresist is exposed to light focussed from a spatial light modulator (SLM), where it interferes with a focussed reference beam to produce a grating, which, unlike that of a dot-matrix pixel, is certainly not a *simple plane grating*. This pixel, instead, contains a complex grating which involves "ray-tracing" the view which the viewer's eye would see from the range of positions it might occupy relative to that individual pixel of the film surface.

Steve McGrew published his "HoloComposer" technique, which has certain similarities to the general principle of the Haines system described, and the techniques of Zebra Imaging Inc. share, to some extent, the principle of ray-tracing the predicted light which would emerge from the individual surface point in the event that a three-dimensional object straddled the surface of the film. Zebra have developed a volume reflection hybrid, which is oriented more towards the display area of holography.

4.9   Zebra Reflection Holograms

Zebra Imaging [8] is today producing many types of hologram, including some which operate on a horizontal turntable basis and are able to represent the bird's eye preview of a scene such as a planned building site, an existing city or a commercial factory estate.

The chosen recording material is a film which, unlike the Haines system designed to originate embossed holography in transmission format, is capable of volume hologram recording, so the result is a reflection hologram. Again, the individual pixel is the location of a ray-tracing prediction of the appearance of that position, as an $x-y$ coordinate component, of a plane within the depth of the 3D scene which is the subject of the image. Because the grating detail is calculated separately for every individual pixel, it is feasible to produce multiple panels which tessellate to produce a large display of massive dimensions – there follows naturally a discussion of the feasibility of architectural displays illuminated by sunlight and my personal ambition of vast holographic sundials in the guise of digital clocks moves nearer! This

Figure 4.9 "Digital Head" by Zebra.

macroscopic concept of holography is the one undisputed advantage of such digital systems over conventional, analogue "entire message" holography, as is re-iterated below by my co-author.

The Zebra method operates in full-colour reflection mode and the gradual reduction in pixel size will result in obvious resolution improvement at the expense of the time taken to record the image via its *x–y* stepping process. Large 3D images are made in 60 cm by 60 cm tiled increments. The first really large image created at Zebra, which propelled the company to sudden fame, was a digital colour image of Ford's P2000 Prodigy concept car, in which ten such hologram tiles made up one very large computer-generated colour reflection hologram. This was unveiled by none other than the United States President Bill Clinton at the 1998 Detroit Motor Show, to much acclaim.

My co-author recorded an amazing image, "Digital Head" at Zebra (Figure 4.9), which he describes thus:

Upon first seeing the completed hologram, I was amazed because I had never seen anything like it in art or holographic science. The hologram was based upon head scans which were made in the UK by Cyber Site Ltd, in Hayes, Middlesex, and a background (rear-plane) photographic image I made, taken of the Arizona desert at sunset. This was a fitting background scene, because the actual hologram was to be made in Texas. For me, the importance of "Digital Head" is best understood against the backdrop of twenty-two years' endurance with the traditional darkroom, starting with the Royal College of Art summer shows in London 1980, where I exhibited fourteen 30 × 40 cm multi-coloured reflection holograms, going on to the very precise intensive monochromatic work made with ruby pulsed portraiture and the large-format display holograms, around which I have since built my reputation. "Digital Head" is the first in a new generation of holograms made possible by the good work of Zebra Imaging. At first, it actually made me feel nauseous, but intuitively I knew this was an image that a new generation of artists could relate to, and one with which I feel deeply contented. Its content is not deeply thought-provoking, spiritually embodying, philosophical or prophetic, but its image power can be measured in megatons. It is grotesquely ugly, but also insanely funny, uniquely original, totally uncommercial and utterly banal. Like the head of the great "Oz" projecting before Dorothy in *The Wizard of Oz*, your eyes will not be able to resist looking behind its hollow head again and again. (M. R.)

Notes

1 van Renesse, R.L. (2004) *Optical Document Security*, 3rd edition. Artech House Publishers (ISBN: 978-1580532587).

2 OVD Kinegram AG, Zachlerweg 12, CH-6301, Zug Switzerland.

3 Wavefront Technologies Inc. 15127 Garfield Ave. Unit B, Paramount, CA 90723 www. wft.bz

4 Lee, R.A. (1991) "Pixelgram: An application of electron-beam lithography for the security printing industry," in *Proceeedings of the SPIE*, 1509, (Ed.) W. F. Fagan.

5 De la Rue Holographics Ltd www.delarue.com

6 Opsec Security Ltd www.opsecsecurity.com

7 Haines, K.A. (1996) "Development of embossed holograms", *Proceedings of the SPIE*, 2652, Practical Holography X, 45.

8 Zebra Imaging Inc. www.zebraimaging.com

5

Recording Materials for Holography

5.1 Silver Halide Recording Materials

The silver halide materials chosen for holography are essentially gelatin-based suspensions which are traditionally known as "emulsions", although technically defying the definition of "emulsion".

These are, in the case of holographic recording materials, definitively based upon silver bromide, although silver chloride is also well known as a really useful photosensitive material, which is used extensively in the emulsions used for photographic papers.

Silver chloride has had no commercial use in the holography arena to date, principally for reasons of grain size and resolution. My early experiments at 3M Research with AgCl emulsions were inspired by the theoretical "colourless" (white) nature of the silver chloride, as compared to the yellow silver bromide crystal, as well as the crystal habit possibilities, but the chloride is relatively soluble in relation to silver bromide and there is real difficulty during emulsion making in controlling crystal growth as a result; the solubility product of AgCl is 1.8×10^{-10} as compared to AgBr at 5×10^{-13} at 25°.

Jeff Blyth [1] has recently established a route by which silver chloride can function in the holographic process by the use of a "diffusion" process, where he has coated and dried a gelatin layer, and treated this layer, after drying, successively with hardener and then soluble silver. After drying again, Jeff finally allows a chloride solution containing dye sensitiser to diffuse into the hardened silver matrix; thus controlling grain growth by time constraint (an example is shown later in Figure 5.7).

The third member of the silver halide family has a significantly lower solubility product 8.5×10^{-17} and, as a result, iodide ions cannot exist in solution whilst the other silver halides are present. Figure 5.1 shows the effect on electrode millivoltage as silver nitrate solution is titrated gradually into mixtures of ammonium bromide and ammonium iodide in a dilute gelatin solution. In each mixture of halides of varying proportions, A, B and C, the electrode voltage is suppressed by the presence of iodide in solution until the excess iodide is converted into silver iodide. At that point, the silver-ion specific electrode voltage increases to a level reflecting the presence of excess bromide. At the point of neutrality, where silver ion concentration moves into excess above total halide, the electrode potential rises sharply. The presence of a silver halide solvent such as ammonia, which is common in the production of conventional photo emulsions, has the effect, shown in the dotted curve, of reducing the electrode millivoltage in accordance with the dissolution of some proportion of the silver halide present in the vessel.

The Hologram: Principles and Techniques, First Edition. Martin J. Richardson and John D. Wiltshire.
© 2018 John Wiley and Sons Ltd. Published 2018 by John Wiley & Sons Ltd.
Companion website: www.wiley.com/go/richardson/holograms

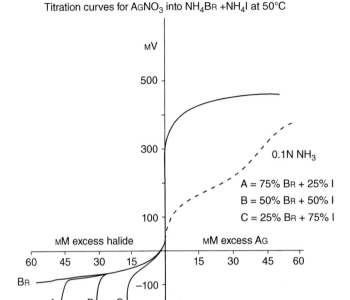

Figure 5.1 Titration electrode potential.

5.2 Preparation of Silver Bromide Crystals

The essence of the preparation of silver halide is the double decomposition reaction between silver nitrate solution and a soluble halide such as sodium, potassium or ammonium bromide.

$$Ag^+NO_3^- + NH_4^+Br^- = \downarrow Ag^+Br^- + NH_4^+NO_3^-$$

Figure 5.2 shows an emulsion growth experiment with a suitable silver and bromide addition to gelatin solution. AgBr was being grown by pAg-controlled jetting to produce a pure bromide emulsion. In the experiment, small samples were taken for real-time crystal size assessment by spectrophotometric (scatter) means.

The plot shows the crystal size increasing uniformly as the precipitation proceeds. However, it is immediately noticeable that the grain size at time zero extrapolates back to a grain size which is already unsuitable for the "ultra-fine" definition to which we aspire for modern holography.

The plot tends to indicate that, without special precautions being taken to reduce the solubility product of silver bromide, the original nucleation step produces crystals which are arguably already relatively large in terms of the ultra-fine-grain holographic products that enable us to make the best holographic recordings. This curve applies to a pure bromide emulsion, so the requirement to incorporate iodide in the emulsion is welcome as it influences grain size, to some extent, in a favourable way.

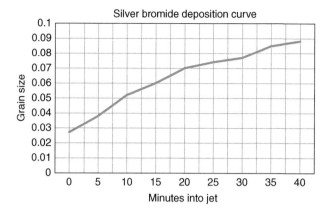

Figure 5.2 Silver halide grain growth plot.

Reflection holography was always, of course, regarded as far more demanding for recording materials than transmission holography; an observation dating back to the time of Lippmann, but in the modern era it has become increasingly evident that the use of first class high-specification materials, even for volume transmission masters, as now used in conventional embossed holography, can result in genuinely perceptible improvements in the quality of the final embossed product.

Traditional embossed holography involves the optical transfer of holographic images from silver halide masters into blue-sensitive photoresist. Any major change of wavelength between such generations results in serious optical aberrations in the holographic image and it is therefore a clear advantage to remain consistent in laser wavelength throughout the origination process, say with 442 nm helium–cadmium or 458 nm argon ion (or d.p.s.s.) lasers of similar wavelengths. The implication is that, in any case, *we require silver halide plates which are compatible with blue exposure.*

This introduces the great difficulty that, in order to prevent scatter and noise in the hologram master, the inclusion of silver bromide grains in excess of ~25 nm diameter is precluded.

Of course, it follows, as we shall see later in the section, that there is a straightforward relationship between grain size and photo speed, so the necessary reduction in grain size has the generally undesirable effect of increasing exposure times in the mastering process.

However, the sole arbiter as regards the relevant qualities of mastering materials *must* be the diffraction efficiency and low-noise capability (signal-to-noise ratio) of the hologram recording plate. Thus, holography has set a very demanding new challenge to the manufacturers of photographic recording materials.

5.3 The Miraculous Photographic Application of Gelatin

When my earliest interest in chemistry, which took the form of making explosives, resulted in the rather distressing removal of the wall tiles from his mother's kitchen, my friend John Trayhorn and I were strongly encouraged to find a new hobby, and we turned our interest to photography.

Making our first silver chloride recording plates in 1964, at the age of eleven, we figured that we could suspend our silver chloride on a glass plate with Vaseline! We did not understand the need for sensitisation at all, and the question of allowing chemical access to the developer apparently escaped us. This is disappointing, because the use of gelatin was discovered 107 years earlier by Richard Leach Maddox in 1871 as a basis for the suspension of silver halide, and the pairing of aqueous gelatin with silver salts has been a literally *miraculous* partnership – gelatin first interacts favourably with silver halide in regard to restraint, or general control, of crystal growth, then produces a relatively dimensionally stable layer, which, when treated with the correct anti-bacterial reagents, provides an excellent lifetime for the coated film. Finally, it permits the desired interaction with aqueous processing chemicals.

Gelatin is a natural product derived from collagen, sourced from animal hide and bone. Because of its source, gelatin also contains matter that could be regarded as impurities, but which, especially in the earlier days of emulsion-making, was responsible for significantly improving the sensitivity of the layer to light.

After completion of film manufacture, and following exposure of the stable layer to light in such a way as to record an image, gelatin allows processing by aqueous chemical solutions, as previously described, which are able to modify the exposed silver halide crystals in suspension in such a way as to permit development, fixing, bleaching, etc., without destructive effect upon the assembled layer.

5.4 Why Has it Taken so Long to Arrive at Today's Excellent Standard of Recording Materials for Holography?

For the holographer, it has been a very frustrating experience waiting for materials to be produced by the traditional silver halide manufacturers of a quality which is capable of fulfilling the potential holography offers as a science for full-colour, three-dimensional imaging. This frustration was exacerbated in the West by the apparent demonstration by Eastern Bloc scientists of the exceptional potential of silver halide for hologram recording, which appeared not to be replicated in Western materials for many years. In fairness to the established photographic suppliers, it was expected in the West to receive a box of plates which were all identical, had an extended shelf life and consistent properties from batch to batch. In practice, Agfa Gevaert went to some lengths to keep regular customers happy by matching customers' specific needs with the output of their factory.

The preparation of plates for the Russian holograms was certainly not based upon such formal commercial principles. We mention later in this chapter Jeff Blyth's successful attempts to produce recording layers by real-time "diffusion coating" to produce small crystals, and such principles were available to the Russian holographers, but realistically not to established Western production units, such as Agfa and Kodak.

There is a long history of photographic film manufacture, which sets useful precedents for the production of films and plate coatings for holography, but *the demands of holography are considerably different from previous photographic products* – far in excess of the requirements for the successful commercial application for microfilm, for example.

It has always been the case that there was an art in growing the desired crystal size and distribution in an emulsion to suit the precise imaging requirements for each application. Such requirements are well described in George Duffin's book *Photographic Emulsion Chemistry* [2] and Mees and James's *The Theory of the Photographic Process* [3].

For example, in an x-ray emulsion, where the very highest film speed is of the essence, it was traditional to have an initial precipitation stage where crystal nuclei were formed. These crystals are intimately associated with gelatin adsorbed at their surface – they are not silver halide crystals "floating" or electrostatically suspended in a gelatin solution (i.e. they are not an emulsion).

At the conclusion of initial precipitation, additional gelatin might be added, to be followed by a "ripening" process which might include ammoniacal solutions or other growth agents to *encourage* crystal growth. Generally, an environment rich in bromide gives rise to growth as a result of complex ions such as:

$$AgBr + Br^- = [AgBr_2]^-$$
$$[AgBr_2]^- + Br^- = [AgBr_3]^{2-}$$

These ions have the general effect of encouraging solubility and crystal growth. *Ostwald ripening* is the name generally given to the process where minute high-energy crystals tend to selectively dissolve in favour of depositing their content upon larger crystals of the same composition, as suggested by Wilhelm Ostwald in 1896.

5.5 Controlled Growth Emulsions

My own work at 3M Photo Research Harlow was directed at researching the modernisation of the silver halide film-manufacturing process in such a way as to replace time-consuming and variable multiple growth phases with a single precipitation by double-jet-controlled precipitation with pumps controlled by drivers receiving electrode feedback of pAg in the reaction vessel.

The pAg of a silver solution is a useful concentration parameter analogous with pH. By definition, pAg is the base 10 logarithm of the reciprocal of the concentration of the silver ion in solution.

$$pH = Logarithm\ 1/[H^+]$$
$$pAg = Logarithm\ 1/[Ag^+]$$

I am proud to say that, as 3M Research exploratory technician of this new deposition method, my colleagues labelled me "Mr Controlled pAg" at Harlow in the late 70s. This "controlled pAg" method was seen, at that time, as the modern answer to crystal growth control.

By the use of these controlled pumping techniques, as indicated in Figure 5.3, I was able to produce pure bromide emulsions which, after running down the corridor to gold coat them *and* bring them under scrutiny of our scanning electron microscope (SEM), contained crystals of a uniform consistency (*monodisperse*) which brought tears to the eyes in terms of aesthetic beauty – but, ironically, these were of little value in photographic terms.

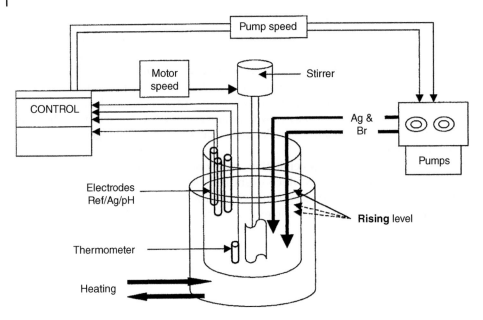

Figure 5.3 Schematic of the electrode-controlled deposition mechanism.

Making beautiful crystals by electrode-controlled deposition is one thing, but it is not a simple matter to replace a film that has existed successfully for years, and 3M's Type R x-ray product had complex properties as regards its ability to display weak shadows, which are, more than ever in the modern era, such a vital part of x-ray interpretation and diagnosis. The means of producing such an emulsion was developed over many years of empirical research. In general, fast emulsions were made by sequences of double-jetting silver and bromide ions into gelatin solutions, where deposition processes were interspersed with "ripening" phases in which crystal growth was encouraged by the Ostwald ripening technique, where minute crystallites effectively dissolve and their content is re-deposited on the larger crystals already present.

Some conventional production processes for certain large-grain (fast) emulsions involved "single-jet" precipitations, where, for example, a vessel containing gelatin rich in bromide, received a rapid addition of silver nitrate solution. In a single-jet precipitation, however, the pAg or silver concentration is changing vastly throughout such a deposition; the effect on crystal habit is incredibly complex, because pAg is influential upon the tendency of silver bromide to produce basic crystal shapes of the cubic or octahedral alternatives; such polar options lead also to hybrids of various types when variations in the conditions of precipitation occur during the procedure. Further, if we are encouraging Ostwald ripening in the digestion stage, the opportunity to create a broad distribution of grain size is particularly welcome. Small nuclei are produced which are later "ripened" at elevated temperatures to produce larger crystals. Silver halide solvents such as ammonia are also used to encourage an environment for growth. The proportion of gelatin is vital to the growth processes and it was often the case that additional gelatin was added in the bromide solution or at the ripening stage.

The photograph of me at work in my conventional emulsion laboratory in Figure 5.4 shows the very basic reservoir vessels that were used in "double-jet" precipitation, with

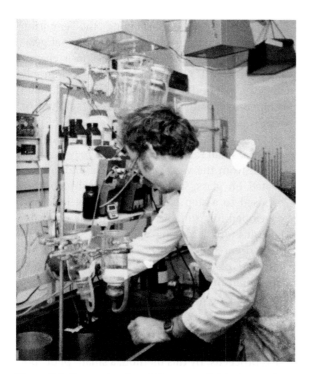

Figure 5.4 Conventional emulsion research laboratory.

capillary glass jets to deliver reagents over a specified time period before the electrode-controlled pumping method was adopted. Obviously in this case, the effect of the reducing levels of reagents in the reservoir meant that the rate of gravity-fed delivery reduced throughout the precipitation.

These are redundant proprietary processes which do not concern us, other than to observe that, from the growth plot in Figure 5.2, such deposition onto existing nuclei is not something we can tolerate in the holographic emulsion area, since the nuclei seen in that precipitation are already of the order of the maximum final grain size required for acceptable scatter in blue-sensitive holographic emulsion.

It is true to say that, in routine silver halide preparation, the visual experience as the precipitation proceeds is the early onset of a milky appearance to the reaction vessel as silver halide crystals begin to grow; but the process for holographic emulsion is quite different, because the solution of gelatin remains quite clear throughout deposition – a sign of opalescence would be regarded as catastrophic in a holographic emulsion.

Once the crystal growth stage is complete, stabilisers are sometimes added, and the emulsion is "washed" to remove the excess electrolytes which are the waste products of the double decomposition precipitation reaction. These are relatively concentrated salt solutions (reagent solutions are typically 1.0 Molar). Apart from other considerations, these would crystallise within the coated layer if left present in the emulsion at the coating stage. Methods of "washing" the emulsion range from the historic method of "noodling" the soft gel into a spaghetti-like condition to be flooded with fresh water, through salt or solvent precipitation or flocculation, to the modern alternative of micro-filtration.

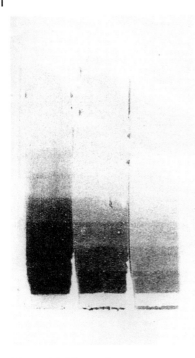

Figure 5.5 "Wet test".

After washing, re-dispersing the emulsion provides an opportunity to add additional gelatin and polymers before the emulsion is chemically sensitised. Conventionally, this commonly involves a period of gentle stirring at an elevated temperature with the addition of sulphur or reduction sensitisers in addition to salts of gold metal or other sensitising species. During this process, small samples of emulsion may be withdrawn from the vessel on a spatula and crudely coated on glass. The gelatin is set by cooling, exposed to a step wedge exposure of light and developed in a suitable developer. The emulsion contains no hardening agent and so must be treated carefully (with cool developer etc.) but the glass plate will provide "real-time" information about the progress of sensitisation, as shown in Figure 5.5. We can see from right to left in the figure, the developed emulsion exposed to a consistent step-wedge of light, increasing in speed, contrast and finally fog, as the sensitisation process proceeds. By this method, a small "pilot" pot of digesting emulsion can be brought up to temperature in advance of the main bulk of the emulsion in the water bath to precede the main pot in digestion time, such that it provides a guideline to the progress of the sensitisation and allows maximum speed to be achieved without risk of digestion fogging.

5.6 Unique Requirements of Holographic Emulsions

With holographic emulsions, where almost any level of crystal growth is regarded as undesirable, the whole question of retaining material at an elevated temperature at any stage in the process is controversial.

This differs so much from ordinary photographic material and has been restrictive for the development of prime holographic materials – even experienced emulsion chemists had no previous experience of such unusual requirements. Even microfilm is far displaced from such demands on resolution and grain size.

So the playing field was effectively levelled and the smaller holography researchers were thus able to produce materials with advantageous qualities over the professional photographic companies.

Sensitisation, nevertheless, is an essential requirement to optimise emulsion speed, so the ability to accommodate this step successfully under advantageous conditions pays a high dividend.

After sensitisation, stabilisers are added to prevent further sensitisation, or stop the ongoing reaction, which would otherwise result in eventual fogging of the emulsion. Another vital addition is anti-bacterial protection, historically in the form of phenol, but now substituted by less toxic compounds, and the emulsion can be cooled and set solid for storage.

Prior to coating, when the emulsion is *gently* melted, spectral sensitisers are added to the emulsion to enhance sensitivity to the relevant wavelengths. The silver halides are *naturally* sensitive to ultra-violet and blue light extending only to about 520 nm in the blue-green part of the spectrum. Each halide has slightly different peak sensitivity and this is related to the respective deep yellow, pale yellow and cream colouration of silver iodide, bromide and chlorides.

Sensitising dye molecules are adsorbed to the crystal surface (with apparent resulting colour change) and, immediately before the coating process, chemicals known as "coating finals" are added. These are wetting agents, plasticisers, gelatin hardeners and the like which enable the production of a suitably durable film or plate coating.

Once again, gelatin has highly favourable chemical properties for our purpose, and the concept of a simple means to control the hardness of the final film layer is a great advantage. Gelatin is a complex protein that has the fortuitous property of "cross-linking", wherein hydrogen bonding can cause its individual chains to become linked, with the effect that the material has unusual rigidity. Even in dilute solution, a network of cross-linked chains will allow gelatin to set; in a more concentrated form, specific hardeners can create an ideal, pliable but solid (compliant) layer for film manufacture. This type of reaction can be brought about by the addition of organic reagents such as aldehydes or ketones, or inorganic ions such as trivalent chromium and aluminium.

I would say that, in the various holographic materials available today, we see a massive range of hardness and wetting properties; these may well influence the ability to create phase holograms with particular reference to deleterious fringe shrinkage, etc.

5.7   Which Parameters Control Emulsion Speed?

In general, larger silver halide crystals will produce an emulsion which is more sensitive than smaller crystals, because the development process is based upon chemical discrimination between adjacent crystals, which either contain a latent image or are not susceptible to development. But there are very important riders to such a generalisation.

The action of a developer, as we shall see later, is to act as a mild reducing agent, with the effect of selecting individual crystals of doped silver bromide, which have been affected by the absorption of photons of light, and selectively reducing their silver ion content to a metallic form which is the black silver with which we are all familiar from the age of black and white photography.

Perfect silver bromide crystals comprise a cubic lattice with an integral period (in the conventional unit) of 5.7 Ångstroms.

However, $1 \text{ Å} = 10^{-10}$ m = 0.1 nm, so we see immediately that our silver halide required crystal size is physically restrictive here, since an ideal silver halide crystal for our purposes of, say, 10 nm grain size, is thus going to contain only a small number of the order of 20 silver ions along the edge of a cubic crystal. We are truly dealing in nanotechnology! Holographic recording is immediately pushing us to the edge of infinitesimal measurements and almost the need to consider individual atoms.

Table 5.1 Atomic weight vs. ionic radius chart.

Element	Atomic weight	vdW Atomic radius (nm)	Typical ionic radius (nm)
Bromine	79.9	0.185	0.196
Iodine	126.1	0.198	0.220

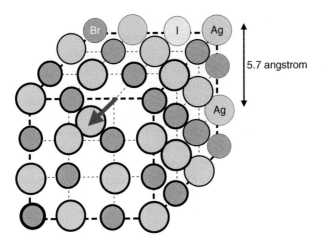

Figure 5.6 Migrant ion creates a Frenkel defect in a doped AgBr crystal.

However, a crystal with a *perfect* cubic lattice, as mentioned previously, is not what is required for high sensitivity. There are suggestions that a *perfect* silver bromide crystal has very little sensitivity to light at all. The mechanism of sensitising the silver halide to increase emulsion speed tends to be assisted by the existence of sites on the surface or the edge of a crystal where the lattice is defective or non-uniform. Such defects may occur for several reasons and the fact that bromide emulsions are made with a low percentage (<10%) of iodide "impurity" is no coincidence.

The larger iodide ions situate themselves in place of bromide with electrostatic equality but cause disruption to surfaces and edges. Although the atomic weights of bromine and iodine vary considerably (see Table 5.1), the ion sizes are not so dissimilar. Van der Waals atomic radii refer to half the distance between atoms in a packed state. These are reproducible, but ion sizes vary according to the environment of an ion. The configuration of electron shells is complex; the overall estimation of radius is dependent upon adjacent ions and electrostatics.

As shown in Figure 5.6, irregularities in the edges and surfaces of the individual crystals provide centres for chemical activity and photosensitivity. Mixed halide incorporation is just one way to encourage their formation.

Frenkel defects involve the absence of a silver ion from its expected position in a uniform lattice, as shown by the red arrow in Figure 5.6. To retain electrical neutrality of the crystal, the ion, termed *interstitial silver*, finds a nearby position in the crystal. Its localised positive charge is able to attract an electron to form a silver atom. These are therefore known as *traps*.

It is not our place to discuss the fine details here, but the existence of defects of the former "dislocation" type leave vulnerable sites of discontinuity which may present a localised excess of negative charge and encourage chemical interaction with positively charged species.

5.8   Sensitisation

Silver halide crystals in the raw condition do not possess high levels of sensitivity. Sensitisation is of two types:

1) Chemical sensitisation;
2) Spectral sensitisation.

5.8.1   Chemical Sensitisation

This category involves relatively simple chemical action at the surface of the individual emulsion crystal. I will not comment upon the industry myth that the first discovery of sulphur sensitisation was a mysterious, supremely fast emulsion, which occurred in an early production environment wherein management investigation revealed that a production operative had dropped his egg and onion sandwiches into the emulsion vat!

However, in the age where component gelatins were manufactured in such a way as to contain traces of active impurities, it meant that the process of simply heating the stirred emulsion in melted condition resulted, in some cases, in a significant increase in the sensitivity of the emulsion.

Later in the history of photographic manufacturing, inert gelatins were introduced, so the process of sensitisation became a separate and more controllable process.

Sulphur sensitisation involves the deposition of infinitesimal quantities of silver sulphide on the surface of the crystal. This is a "sensitivity speck" that has the ability to facilitate the amplification process which defines the function of silver halide as a recording material.

- The crystal defect offers a site for sensitisation.
- Sulphur and precious metal sensitisation produces a number of centres on each crystal which are ultra-sensitive to light.
- Absorption of light energy (of the order of four photons) at these sites results in the formation of photoelectrons, which produce free silver atoms.
- Development results in an autocatalytic process which converts the individual crystal to metallic silver. Unexposed crystals are immune to this, provided appropriate developers are used.

5.8.2   Spectral Sensitisation

This refers to the treatment of the basic sensitised silver halide emulsion with dyes which interact with the silver iodobromide crystal in such way as to impart sensitivity to light of specific visible wavelengths, other than the high-energy violet/blue light that inherently reacts with silver iodobromide, which is pale yellow in colour and thus naturally tends to absorb light of such wavelengths.

Light which causes photolytic action on the recording material is called *actinic light.* Without additional dye compounds which will absorb light being added to the silver halide emulsion, a very high proportion of yellow-green, yellow, orange and red light incident upon a film layer will tend to pass through the layer without any effect on the silver iodobromide. In fact, this is the basis of the principle of "safe lighting", which is discussed in Chapter 7; we can generally find wavelength zones which have little or no photolytic effect on even relatively fast recording materials. These are used conservatively at low levels of intensity to allow us to handle and manipulate the recording material without contributing spurious exposure which would otherwise be deleterious to the required recording.

There are groups of dye compounds which have very favourable properties in terms of their ability to absorb photons of light of a specific wavelength and to propagate the transfer of photoelectrons into the silver halide crystal. Such dyes attach themselves to the crystal surface in highly specific ways: stacking planar molecules along the crystal surface. Dyes are only effective when attached in this way to the crystals of silver halide; left unattached within the gelatin layer, they can actively *reduce* the emulsion sensitivity.

Because the adsorption of sensitising dye is a superficial process, there is an interesting consequence of the massive surface area presented by the very small crystals of a holographic emulsion. So, for example, a fast x-ray emulsion could typically have a grain size greater than one micron, so might require of the order of 100 times less sensitising dye per mole of silver than a 10 nm ultra-fine-grain holographic emulsion to saturate the crystal surface at the same level.

In practical (and cost) terms, there is a considerable problem in delivering such a quantity of dye to the gelatin solution, since non-sulphonic cyanine-type dyes are often unsuitable for aqueous solution, whereas gelatin is simultaneously susceptible to precipitation by solvent. Again, we see the unique resolution requirements of holography presenting wholly new technical problems to the manufacturer.

5.9   Developer Restrictions

Great selectivity of the properties of developers is required for holography. The very highest resolution is required from the emulsion, whose contents will, in general, be required to be converted to a phase hologram.

But the minute crystals represent a high-energy state which will tend to create problems of solubility in typical developers where bromide, for example, is present.

Lithographic developers involve "infectious development", which causes crystals adjacent to the exposed crystal to develop; in holographic applications, we need to avoid such occurrences to preserve the integrity of the individual crystal and thus improve the resolution of the diffractive structure.

Because of the need for extremely high image resolution to record hologram fringes, the holographic emulsions necessarily comprise the smallest grains of silver halide that can be made. In the early 1980s, my discussions with manufacturers revealed that it was generally felt that ~35 nm grain size was all that was required to achieve perfect red holographic recordings. However, I think the point is missed here that the grain size of the original emulsion is not necessarily the limitation on the final resolution of the hologram. In fact, the bleaching process tends to define the microstructure of the final hologram, and it is possible for the various bleach formulations either to reduce or increase the effective grain size and thus the resolution of the microstructure recording.

In this regard, the smaller the initial mean grain size, the more freedom we have with the bleach formulation, and the preferred "re-halogenating" bleaches described in Chapter 6, which are capable of high diffraction efficiency, are suitable for use without undue scatter and noise effects.

5.10   The Coated Layer

It is well known that in photography, the ability to coat silver halide emulsions in *a single-step coating process* in a stack of discrete layers with differing chromatic sensitivities has allowed colour recording of photographic images to proceed at astonishing levels of quality in the "slide/transparency" and the "colour negative" modes of operation.

To date, it has been advantageous to produce a single layer for holography materials wherein there is effectively a mixture of spectral sensitivity spread through the crystals of the emulsion. I am not aware whether sophisticated research has been carried out to establish the details of the configuration of the dye/crystal relationship. In view of "j-band" type phenomena, it would be surprising if optimised sensitisation could be achieved by the addition of multiple dyes in random sequence to a single emulsion.

However, the multi-layer configuration routinely used in colour photography has not been used commercially because the thickness of the emulsion for volume holograms is, *in its own right*, vitally important. For the main part, silver halide functions, in all cases, as a Bragg volume hologram, as discussed in Chapter 2, even when used in transmission mode. For this reason, each component microstructure (each colour recording) requires the maximum possible volume of emulsion (*number of fringe layers*) to provide the optimal diffraction efficiency. Theoretical calculation indicates that emulsion coatings of, say, 20 μ thickness are required for maximal D.E. at the typical expected levels of index modulation, Δn. But such thickness is not realistic for flexible films with suitable physical properties, so, in practice, commercial silver halides have typically been coated at 8 μ or less.

It is therefore arguably more satisfactory to have all of the volume Lippmann fringe structures in these films, associated with each component (RGB), extend throughout the whole available depth than to confine them to, say, one-third of the thickness, even in the event that the data storage naturally becomes somewhat confused and overcrowded.

It is not necessarily the case that holographic recording emulsions must always be coated upon ordinary pre-subbed glass plates or film. My co-author has recently been working with Jeff Blyth at De Montfort University. Jeff has produced layers of gelatin which were coated on the *inside surface* of glass objects such as wine glasses and filter funnels. As described previously, the photosensitive layer was based on silver chloride made by a diffusion technique into a hardened layer of gelatin. Exposure by low-power lasers has allowed the production of "Denisyuk" holograms where a small model is placed within the coated glass vessel and exposed to laser light. As Jeff succinctly points out in his RPS article [1], the wine glass acts as its own reservoir for both the coating and

Figure 5.7 Wine glass hologram.

processing solutions. The result of illuminating the hologram with white light is a realistic reconstruction of the image within the glass, as shown in Figure 5.7.

5.11 The Non-typical Use of Silver Halides for Holography

The silver halide manufacturers generally found it difficult to apply their exceptional photographic expertise to holography.

My own experimental emulsions made at 3M Research in 1982 defined the gulf between holographic and conventional emulsions when, after leaving new holographic emulsion candidates for trial coatings, I received an urgent call from my colleague, Peter Pavitt, the highly experienced 3M coating technologist, with the message, "I think you've forgotten to put silver in this emulsion – it's completely transparent!"

This illustrates just how exceptional is the difference in grain size between holographic and conventional photographic materials. Nevertheless, Agfa Gevaert supported the holography industry in Europe throughout its formative years, and it was Kodak who produced early materials which allowed Leith and Upatnieks to forge their incredible developments in holographic technology. Ilford Ltd, in the 1980s and 1990s, developed materials for Applied Holographics' use of ruby pulse lasers and also blue/green materials that, as a result of a small reduction in mean grain size and improvement in grain size distribution, exceeded the quality of previous commercial materials for embossing origination work, especially with regard to blue scatter.

The revolutionary step for my own purposes was the work of Jeff Blyth and Richard Birenheide to create the HRT emulsions. They made possible a really significant step in my own progress with mastering for embossed holography, and since Mike Medora took control of that process, further refinements have been made for a material which has magnificent properties including incredible process latitude.

Over the years I have made holograms on my own 3M film, Agfa, Kodak, Ilford, Fuji, ORWO, Minolta, Slavich, Harman and Colour Holographic HRT. The latter emulsion has given absolutely exceptional results in both embossing master work with blue lasers in transmission and in full-colour reflection work. Today, other holographers around the world are making exceptionally high-quality full-colour Denisyuk reflection work both with the Colour Holographic and with Yves Gentet's Ultimate products.

The essential difference between the materials which create exceptional image quality is the emulsion grain size. In the gradual diminution of grain size from the earlier Agfa materials with a mean grain size in excess of 35 nm, through the small reduction in the Ilford materials of the late 1980s at around 30 nm, we have seen, in recent years, the crystal size reduced below 20 nm for the first time. As well as having a really small grain size, the Colour Holographic BBV materials are highly reproducible, but the particular technology used appears to resist scale-up to levels considered viable by the major manufacturers.

There is a need to ensure that holographic emulsions designed for "volume holography" are of optimal coating thickness to provide an appropriate number of layers of index-modulated material, whilst retaining minimal levels of scatter and transparency. The need for a traditional "supercoat" of hard gelatin provides an ongoing topic of discussion in a layer which would benefit from a maximum thickness of active coating, but is not highly sensitive to stress marks, as would be the case for a fast emulsion, so why not use all the available layer thickness for active material?

Photographic materials are required almost universally to show maximum recording speed and to reach the required silver density. Contrast potential is frequently adjusted within the emulsion-making process in order to achieve materials which are suited to individual process requirements; for example, in the manufacture of x-ray film, there are critical details of image information (shadows) which manifest themselves as minor density variations in mid-tonal areas of the dynamic range of the emulsion. Conversely, in a lithographic material, the film, in association with special developer formulation, is designed to produce a "digital" response, where adjacent zones permit, ideally, either 100% or zero density to facilitate half-tone work required for print applications.

In a holographic emulsion, we automatically require a "high-contrast, high-resolution" emulsion. It is often stated that interference fringes in a standing wave of the type we wish to record in holography are of a sine or cosine profile, and that silver halide is capable of recording such infinitesimal detail. This is an example of the error in assuming that any limitation of the feasible maximum resolution is acceptable for holography. For example, 1000 l/mm resolution is a typical microfilm standard, which would apparently allow the recording of a basic fringe structure, but it will not facilitate a cosine profile such as is required in volume holography.

Interestingly, the effect of "fog" from any source whatsoever has quite a different effect in holography to the photographic counterpart:

1) The emulsion will be bleached, with the effect that fog is converted to "noise". This "noise" is closely associated, especially when blue lasers are involved, with the scatter resulting from the general phenomenon of Rayleigh scatter.

2) Figure 5.8 shows the comparison of scatter effects for the three common blue laser lines in typical use in embossed holography. It demonstrates with clarity the reason why emulsion producers were able to relax in the aspiration for grain size reduction in the case where the ubiquitous HeNe laser was considered the light source – it makes little difference to the 633 nm scatter level when the 20 nm and 30 nm grain sizes are compared.

3) The increase in minimum density (D_{min}) is directly responsible for a reduction in Δn, the index modulation, in the same way as a reduction in D_{max} affects the index modulation simply by reducing its upper constraint. The "diffusion-transfer" bleaches are effectively transporting material from zones of low density to zones of high density; for this reason, it is probably the gulf between these extremes, rather than their absolute density, which is critical. The details of processing silver halide film are discussed in Chapter 6.

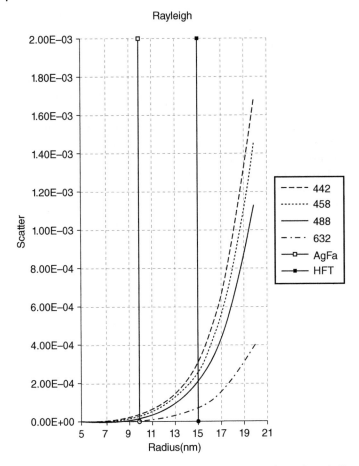

Figure 5.8 Rayleigh scatter as a function of grain size and wavelength. Figure courtesy of Alex Cabral.

5.12 Photopolymer

When introduced to the Du Pont photopolymer in 1987 during the visit of Du Pont scientists Dalen Keys and Bill Smothers to England, it is true to say that I was absolutely astonished (and a little confused) after setting up a contact copy camera in the studio at Applied Holographics at Braxted Park.

There, Simon Brown and I watched in amazement as live bands of interference ran across the assembly of film and a planar coins model. This "movement" appeared contrary to everything I had learnt about hologram recording prior to that experience.

At that time, the material was sensitive to blue (for example, 488 nm) and blue-green (for example, 514 nm) light only and required subsequent baking to stabilise and improve the image.

Today, using photopolymer regularly, I recognise that the reflection of laser light from the object was interfering with the light reflected from the growing Lippmann hologram of same subject matter, and that every index *improvement* in the hologram microstructure represented phase change that resulted in mismatch with the phase of laser light reflected from the surface of the actual coins. Thus, the *growth* of a diffractive microstructure is responsible for visible fringes of interference rushing across the surface of the film during the imaging process.

From the point of view of Lippmann or Gabor, one can imagine that the concept of a material actively recording their predicted interference pattern *in real time* would seem a fantasy of the highest order, but this concept has today offered us a whole new route for the realisation of holographic recording.

For our purposes, there is a considerable difference in the imaging action which occurs in silver halide and photopolymer, because, quite simply, we can expose photographic/holographic information into silver halide in consecutive exposures without any tangible effect until such time as the photosensitive material is submerged in developer. Conversely, in photopolymer, we witness the spontaneous changes within the layer, which result in the production of differential refractive index zones capable of diffractive effects through the direct formation of a phase hologram.

There is a considerable difference in "photo speed" between the silver halides previously mentioned and the photopolymer type of reaction, which significantly affects the way in which these materials are handled. Of course, this provides a great advantage in terms of safe light conditions for the studio, but laser power and exposure times, and their relationship with optical system stability, are adversely affected. In general, we may be looking at a difference of the order of fifty times more energy to render the necessary changes to create an image.

Photopolymers have been used in specialist photo-printing processes for many years, and many formulations which are familiar in the polymer context have been tested, such as:

- Acrylates
- Acrylamides
- PMMA
- PVA

Researchers have recognised the need to achieve increasingly high resolution in the basic formulations in order to include volume holography in the repertoire of photopolymer, and Polaroid, Du Pont, Dai Nippon, Techsun, Xetos and Polygrama have all produced realistic candidate products. The new Bayer HX photopolymer by Covestro is a polyurethane-based material [4].

Although the natural sensitivity of such monomers tends to be principally in the UV region, modern chemistry allows for interaction with visible light; indeed, in the case of the Covestro (Bayer) HX photopolymer, we see a full range of spectral sensitivity, including a remarkable level of activity at the red end of the spectrum.

In principle, the photopolymers comprise a binder polymer, with a monomer content, which can be activated by a photo-initiator to allow visible radiation to precipitate polymerisation of the monomer component. This process is thus able to promote migration of monomer species in accordance with microscopic light intensity variations, which facilitates refractive index and density changes in the layer volume in accordance with the standing wave associated with a holographic image.

These index modulations within a "thick" layer are, in general, locked into place and, in some cases, amplified or exacerbated in the various corporate formulations by incoherent (flood) illumination, UV exposure, heat application or chemical influence.

Saturation with any of these effects allows the elimination of further sensitivity and the permanence of the holographic image.

The index-modulated microstructure is able to induce phase changes, in accordance with the Bragg condition, in incident light at the reconstruction stage, thus enabling such photopolymers to be ideal carriers for reflection volume holograms.

As we have discussed previously, the existence of silver halide crystals of finite dimensions in conventional silver-based recording materials is problematic for that genre of films, both in terms of resolution limitations and of susceptibility towards Rayleigh scatter. However, in principle, the photopolymers can be regarded as grainless.

Of course, the "real-time" chemical and physical changes within the layer during exposure call for a different approach to the handling of silver halide materials, where a number of "latent images" can be posted before any significant chemical change occurs at the processing stage. In photopolymer, the ongoing exhaustion of photo-initiator and monomer material in the layer can, quite simply, result in the fact that, after the initial exposure of the material, a subsequent exposure may fall upon an inert layer exhausted of active material.

So, for example, when silver halide is subjected to consecutive exposures of light of three wavelengths, provided these are not unduly excessive and are relatively well balanced with regard to relative spectral sensitivity, the development process can afterwards be controlled to enable all three fringe structures to form satisfactorily. The use of photopolymer in similar circumstances may lead to a result which is biased in favour of the first exposure grating at the expense of later exposures, since monomer is being used up as the exposure proceeds; subsequent exposure requires that the source material has not been exhausted previously.

During simultaneous exposure to multiple wavelengths, there is a level of aggregation of energy whereby mutual latensification occurs, and in some circumstances similar latensification can be achieved by incoherent (white) light.

The serious inhibitor to polymerisation is oxygen. For this reason, the photopolymer products are generally coated as a "sandwich" of layers, as shown in Figure 5.9. The protective film prevents aerial oxidation or oxygenation of the active layer. This protection must continue to be provided during the actual exposure to the holographic standing wave.

The removal of the thin protective film exposes the sticky surface of the active recording layer. Whereas this appears, at first, to the experienced user of other recording materials such as silver halide film to be rather inconvenient, there is, in practice, an important advantage. In view of the relatively low photo speed of the photopolymer material, the need to hold the material in a sufficiently stable position throughout the

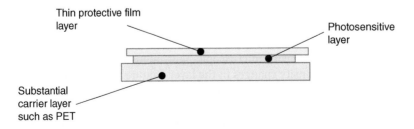

Thin protective film layer

Photosensitive layer

Substantial carrier layer such as PET

Figure 5.9 Typical photopolymer layer assembly.

exposure period becomes a difficult task. This task is assisted by the ability to attach the film to a holding plate in the exposure gate.

Additionally, the exo-thermic nature of the polymerisation process requires that energy is released during exposure and, in effect, the holding plate acts as a heat sink for this purpose. Finally, the adhesion of the smooth surface to the holding plate is sufficient to optically couple the film to the glass and thus eliminate internal reflection from the interface.

Provided the removal of the protective layer is succeeded by lamination to the holding plate, the exposure to aerial oxygen does not create a problem; the actual exposure time is the critical period during which air inhibition of the polymerisation reaction can occur. The same principle applies to any subsequent curing exposure which is used to complete polymerisation to stabilise the layer with its completed holographic image. Such exposures may have a dual purpose; there is a need not only to complete the polymerisation process but to complete the decolourisation of residual sensitising dyes.

5.13   Photoresist

In various industries, such as microelectronics and CD manufacture, photoresists are materials that are used to coat various substrates, with the purpose of providing a barrier layer to allow chemicals to be used selectively to etch the under layer in a pattern dictated by photographic imaging or masking.

But embossed holography has benefitted from the fact that the resolution characteristics of certain resists allow superficial (thin) holographic fringe structures to be imaged into the surface to produce surface-relief gratings.

Dow photoresists [5] were originally produced by Shipley (Rohm and Haas), principally for use in the electronics industry for the production of integrated circuits of very high precision. There are two forms of photoresist for optical imaging: positive and negative systems. In a negative system, the exposed area is rendered insoluble in the developer by the action of blue, violet or UV light; in a positive system, which is generally preferred for holography due to higher resolution capabilities, the action of the light is to promote solubility in the developing solution.

The 436 nm line of the mercury spectrum is the peak sensitivity area of the Shipley S1800 series resists, so that, in embossed holography, the 442 nm line of the helium–cadmium laser is extremely actinic per unit of power. The 458 nm argon line is also effective, as is the 413 nm krypton line, but the sensitivity of the material tails off badly as regards use of the 488 nm argon or 491 nm d.p.s.s. lines. Electron beam exposure can also be used to expose other photoresist types, so that e-beam lithographic forms of holography are also served by materials of this general form.

When coated as a very thin layer of the order of 1.5 μ thick, photoresist can be used for the production of surface-relief transmission holograms as required for the embossing process. The liquid resist is generally coated by spinning at various solvent dilutions and speeds calculated to achieve the required thickness. Since the period of a typical embossed hologram is of the order of 0.7 μ, such a thickness is ideal to store a cosine-profile relief. During the processing of positive resist, there is a dissolution of exposed material, which is necessarily highly controlled in order to avoid the overall thickness reducing beyond the optimal diffraction efficiency.

Professional coatings are available from companies such as Towne Technologies Inc. [6], whose experience and skill enable the user to work with precise, flat coatings of the highest quality, but it is also possible to coat the liquid resist by dip coating, curtain coating or spray coating. The resist is baked after coating to harden and eliminate solvent.

The spray-can electronic circuitry photoresists supplied by CRC Industries [7] can be used to produce coatings which are capable of recording rainbow holograms for experimental purposes with a blue laser. (It is, of course, difficult to achieve a perfectly smooth layer by this method.)

Resist for embossed holography is often coated upon flat soda lime substrate glass with an under layer of red/brown iron oxide, which is able to absorb blue laser light in order to prevent spurious reflection from the glass surfaces. Other, more temporary anti-reflective layers can also be coated on the rear surface of plain glass for this purpose. In the past, I used excellent coatings of this type from Ibsen Photonics [8] on clear glass coated with a dyed gelatin layer on the rear surface which could easily be washed away after exposure and before processing to produce clear optical elements.

Of course, photoresists coated at thicknesses of less than a few microns are not capable of "volume holography", since the elimination of exposed material in the "development" process is via surface dissolution, but these coatings are capable of recording a range of fringe profiles in thin holograms, including *blazed grating*s to a certain degree.

The processing of photoresist for holography is a critical process described in Chapter 6.

5.14 Dichromated Gelatin

Dichromated gelatin (DCG) has played a rather glorious role in holography in the past. As gelatin is such a remarkably inexpensive material, this method, which is capable of very high index modulation and diffraction efficiency over both narrow and wide spectral bandwidths, appeared to have a promising future until the advent of the photopolymer recording materials for holography.

However, there are significant disadvantages to the DCG process, including:

- The growing awareness of the toxicity of chromium ions;
- The need for high levels of laser energy to expose the film;
- The complexity of the processing technique;
- Limited spectral sensitivity;
- The need to hermetically seal the layer to ensure permanence of the holographic image.

As a volume hologram recording material, DCG was responsible for the growth of two aspects of hologram technology which could not be more diverse: jewellery pendants and holographic optical elements (HOEs) in the form of head-up displays (HUDs) for fighter aircraft!

The popularity of decorative pendants for necklaces and watch faces offered an opportunity to get Denisyuk holograms into the public eye. The DCG layer was enclosed between thin glass pates which were sealed together to prevent the encroachment of air or moisture onto the hardened gelatin layer.

Pilkington Avionics HUDs for the F16 Lantirn fighter aircraft navigation system comprised a narrow-band reflective element which was able to allow the pilot a view, focussed at infinity, of the control panel in a single narrow-band wavelength. Unlike a partially silvered screen, a very high proportion of the external natural light passes directly through the optical element, so the pilot's view is not impeded. The projected content of an infra-red visualisation system (night vision) could be included in the projected view.

5.14.1 Principle of Operation of DCG

The inclusion of ammonium or potassium dichromate solution in a gelatin layer results in the potential for *photolytic action, whereby photons of blue or blue-green light can induce transfer of electrons from the dichromate ion $(Cr_2O_7)^{2-}$*. The resulting chromium ions in the Cr (III) state are thus distributed within the volume layer in accordance with the standing wave interference. Local hardening occurs by cross-linking of gelatin chains, and the resulting microscopic zones of hardened gelatin provide contrasting solubility when the layer is then washed.

When the unhardened zones of gelatin preferentially dissolve in the wash, voids occur in the layer which contrast starkly in refractive index with the remaining solid, hardened gelatin. The effect is that the index modulation, Δn, for DCG is higher than other holographic recording layers can achieve – this may be of the order of 0.08.

5.14.2 Practical Experimentation with DCG

The very first consideration when experimenting with DCG is one of safety. Here, we are dealing with the very undesirable Cr (VI) ion, which is carcinogenic and whose negligent distribution is responsible for endless legal suits across the globe. Furthermore, the low photo speed of the DCG system (significantly less sensitive than Bayfol or Du Pont photopolymer) means that blue or green lasers of high power are required to achieve satisfactory exposures. So, whereas the proposition initially appears attractive to the amateur, the health and safety issue is prominent, and protection of the skin and eyes from chemicals and laser light is absolutely vital.

To achieve a suitable working layer, it has been suggested that the application of fixer to unexposed silver halide plates is a good way to produce the required gelatin coating. Proprietary fixer contains hardener, so, as well as removing the silver halide it will further harden the whole layer. This can be a successful technique, but clearly a very expensive and rather wasteful way to produce a plain gelatin layer!

This layer is soaked in a dilute (5%) potassium dichromate solution (not acidic dichromate) for several minutes and then dried in subdued lighting.

A purer route is to coat new, inert photographic gelatin by soaking, say, 50 g in 500 ml of distilled water for half an hour to allow the water to be absorbed to expand the dry gelatin. The solution is warmed, with stirring, to 45° to allow a complete dispersion of the gelatin. There is a balance between adding a small quantity of hardener in the form of 1% formaldehyde or glutaraldehyde to produce a durable layer, and leaving the gelatin soft to allow the photolytic dichromate reaction to have greater effect. At this stage, gelatin can be coated onto a pre-subbed glass plate for subsequent dichromate sensitisation, or, alternatively, the dichromate can be added to sensitise the gelatin prior to coating the layer.

A Meyer bar can be used to produce a good-quality coating at the required thickness. These are wound with wire at a range of gauges and it is possible to apply strips of thin tape to the edges of the plate to increment the layer thickness to the required level, remembering that the wet gel will contract to a fraction (<10%) of the thickness after setting and drying. DCG layers will normally need to be of the order of 10–20 μ to achieve optimal D.E. after processing.

Slavich [9] is one manufacturer which supplies coated gelatin for this purpose. The PFG04 film has a shelf life of one year.

The art of DCG holography is a test of dexterity and every worker develops their own preferences. We can be spurred on by the remarkable D.E. and layer clarity and spectral bandwidth control that some exponents, including some of the professional technical manufacturers of HOEs such as Pilkington and Marconi, are able to achieve.

The prepared dichromated plates should be allowed to stand for some hours after preparation, but their shelf life is limited to perhaps a few days.

Denisyuk single-beam holograms provide the ideal usage scenario. Exposures of about 100 mJ/cm^2 are to be expected at 458 nm and depend upon the details of the gelatin type and the preparation method.

After exposure, the plates are allowed to stand warm for some minutes prior to a water wash (some workers use fixer with hardener) to remove the more soluble gelatin components, and the plates are then dried suddenly by immersion in alcohol solutions of reducing aqueous dilution. The final wash in pure propanol is followed by desiccation. The appearance of the holographic image is spectacular and exciting when the film dries. Manipulation of the above conditions can control image colour, bandwidth and brightness.

The layer must be sandwiched with glass to seal against moisture with UV-cured adhesives. Any minor leak in this seal will admit sufficient moist air to destroy the holographic image over a period of time.

5.15   Photo-thermoplastics

At the opposite end of the scale of photo speed is the only system which can exceed the speed of the silver halide films. This is the photo-thermoplastic material which has been used in non-destructive testing laboratories. This is not only the fastest recording material, but is sensitive to a wide spectral range of visible laser light and has the added advantage that the film or plate can be erased and re-used for many cycles. It is, in effect, a surface-relief recording material for transmission holography; there is no volume capability.

In principle, as shown in Figure 5.10, a conductive under layer is coated on film or plate to act as a ground; upon this is a sub-micron photoconductor. The surface thermoplastic

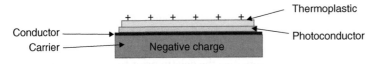

Figure 5.10 Schematic of photo-thermoplastic assembly.

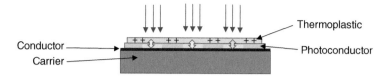

Figure 5.11 Mechanism of photo-thermoplastic exposure.

layer itself acts as an insulator and, prior to exposure, the film is subjected to corona discharge.

The spatial distribution of light intensity in the fringe structure renders a conduction potential in the photoconductor. Thus, local static charge is able to flow and neutralise, as shown in Figure 5.11, in accordance with the image recorded.

The remaining image-wise static charge, when the layer is heated, causes the liquefied thermoplastic to flow into a relief pattern corresponding to the configuration of the charge.

Notes

1 Blyth, J. (2016) "The new Blyth diffusion method for making holograms on glass plates and wine glasses." RPS 3D Imaging and Holography Group.
2 Duffin, G. (1966) *Photographic Emulsion Chemistry*, Focal Press Ltd.
3 Mees, C.E.K. and James, T.H. (1966) *The Theory of the Photographic Process*, 3rd edition. Macmillan (ISBN: 978-0023800405).
4 Dr Friedrich-Karl Bruder *et al.*; http://www.bayer.com/en/bruder.pdfx
5 Shipley Photoresist (www. Dowcorning.com)
6 Towne Technologies Inc. (http://www.townetech.com)
7 CRC Industries Europe NV Touwslagerstraat 1 9240 Zele Belgium www.crcind.com
8 Ibsen Photonics (http://www.Ibsen.com)
9 Slavich (http://www.slavich.com)

6

Processing Techniques

6.1   Processing Chemistry for Silver Halide Materials

As a general principle, holography utilises the time-honoured techniques for processing silver halide. Developer...fixer...bleach are the common vocabulary in the photographic processing room. However, we need to bear in mind the slightly modified objectives of the methods of processing film and plates in the holographic application.

In short, we are ultimately interested only in producing bright, noise-free images in the holograms; this may contrast with the well-known objectives of black and white photography, where image contrast, silver density, freedom from fog, etc. are the parameters under immediate scrutiny.

Whereas the photo emulsions used for holography resemble, to a great extent, those utilised in black and white photography, or lithographic film, the modern emulsions, as previously discussed in Chapter 5, comprise silver bromide grains which are significantly smaller than those in any other application, and the emulsions therefore require special treatment, not least from the point of view of their tendency for solubility during the developing and bleaching processes.

I was fortunate that my career path led me from the preparation and research of silver halide emulsions directly into holography. At the time when I left 3M Photographic Research laboratories at Harlow to join the embryonic Applied Holographics in its original Essex setting, the company's mission was to mass produce reflection holograms, and I was thrown directly into the task of investigating suitable film and processing technology to fall in line with the ruby pulse exposure method which had been identified by Hamish Shearer and Gavin Ross. I was fortunate, in that environment, to meet Jeff Blyth who was working at that time as a consultant to AH, with the effect that we took part in useful discussions that allowed me to tune my knowledge of conventional photographic chemistry with Jeff's specialist subject of holographic processing.

Unlike the pure research environment of my emulsion work at 3M, all of my research and development work in holography has been in the pursuit of commercial solutions to complex problems in real time.

So much data is available in the profound, expert accounts of the application of silver halide technology to holography, as described in publications such as Hans Bjelkhagen's *Silver-Halide Recording Materials for Holography and Their Processing* [1] and the late Graham Saxby's publication *Practical Holography* [2], that research for the aspiring

The Hologram: Principles and Techniques, First Edition. Martin J. Richardson and John D. Wiltshire.
© 2018 John Wiley and Sons Ltd. Published 2018 by John Wiley & Sons Ltd.
Companion website: www.wiley.com/go/richardson/holograms

holographer who wants to make a first simple hologram becomes extensive. We are, perhaps, spoiled for choice in the number of options available.

In this book, therefore, we have attempted to supply ideas of relatively simple working techniques that offer a newcomer to the territory a means to enjoy early practical success and to provide ideas as to areas worthy of further research. At that stage, beginners may wish to explore the more complex chemical issues of silver halide processing. There is unlimited information available nowadays regarding individual chemicals and effects that have been used to advantage in photography, and many of these can be modified for use in our holographic application.

I have had so many calls over the years from frustrated enthusiasts attempting to solve the problem of their failure to produce a first successful holographic image, and it is clear that the remarkably wide range of possible explanations for initial failure to create an image presents a really difficult problem to solve; especially since it is not immediately clear whether failure lies principally in the inter-related areas of laser exposure or processing chemistry.

In this chapter, therefore, we will refer to general principles only; we "cut to the chase" so as to suggest starting points for processing which will yield bright holograms in common materials and suggest simple ideas for optimisation to suit the particular application.

In an ideal world, we might aspire to have a small number of bottles of processing solutions on the laboratory shelf, with unlimited expiry dates, which would provide universal success in terms of diffraction efficiency and signal-to-noise ratio for all hologram recording materials. In some cases, particularly in earlier days, holographers have used proprietary developers such as Kodak D-19b, Ilford PQ and Tetenal Neofin Blue. These chemicals can give perfectly acceptable results and might be very useful in the initial setting-up procedure for proof of principle of a holography project.

Unfortunately, in summarising the methods of processing for the range of materials available, my experience is that the apparently minor differences between the recording materials, which, from my own background of making emulsions, can result from apparently minuscule differences in crystal structure, sensitisation, gelatin hardness, coating weight, etc. can, in fact, be responsible for great differences in optimal processing practice. It is therefore necessary, at the very least, to acknowledge that the holographer must optimise a processing regime for each recording film and for the individual *optical* set-up. If this is acknowledged, the preparation of processing solutions from basic chemical ingredients provides a solid basis for optimisation.

As an extreme example of the need to optimise the processing method for each application, I would point out the fact that the optimal beam ratio for a specific image in a standard H1/H2 transfer camera cannot be defined without knowledge of which recording material, processing method and exposure time is to be used. If we take two fundamentally different recording systems such as silver halide and photoresist, then, for example, the silver halide will require a far higher proportion of object light than the photoresist. I have seen several examples where the transfer of an H1 image into a second generation photoresist required four times less measured object light than the same transfer into silver halide. Moreover, the AgHa hologram itself will require quite different ratios to optimise results in differing processing chemistries.

In holography, the exposure process itself is far more complicated to summarise than it is for photography, especially considering the obvious effect of variable beam ratio between "reference" and "object" light, which may vary significantly from one hologram recording to another. Unlike conventional black and white photography, the development of silver itself does not provide an immediate visual indication of success; hologram efficiency is an indirect derivative of the full processing regime.

We have seen the theoretical representation of both thin and thick grating "fringe structures" as "lines" and "planes" of index modulation within the recording layer. When analysing processing methods, we must keep our attention very firmly fixed upon the fact that, at the microscopic level, silver halide is essentially a granular material; to some extent effectively *digital*, because individual crystals are *either* developable *or* not developable. The classical interpretation of the standing wave of interference as a sinusoidal index modulation of the layer is an extremely demanding target for the recording process.

In Figure 6.1, the schematic represents the standing wave in the classical volume hologram created by two incident waves of green laser light in opposing directions into the recording emulsion in approximate scale comparison with the dimension of silver halide crystals of the order of, say, 20 nm.

The individual crystals of silver bromide are, in fact, relatively well dispersed within the gelatin-based recording layer when the film is made, but unsatisfactory processing can easily lead to the production of agglomerates which may reduce resolution and increase scatter.

The true picture of the microstructure within the silver halide emulsion layer belies the envisaged perception of discrete layers of high and low index, which might well be the case if we examined microscopically the layer of a deposition-coated dielectric mirror, for example.

The granularity of a silver halide hologram illustrates how the particulate nature of the layer results from the finite grain size of the original silver halide crystal. The photomicrograph shown in Figure 6.2 by Harman Technologies Ltd [3] shows silver halide grains in a volume hologram. The distribution of grains amounts to a classical fringe structure in the form of a *granular* sinusoidal modulation of index, reduced to the form of a mean zonal integral value of index, in planes recorded in the layer.

Figure 6.1 Density profile vs. crystal population.

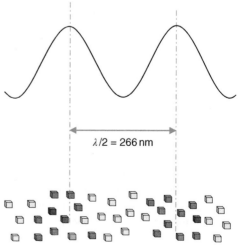

$\lambda/2 = 266\,\mathrm{nm}$

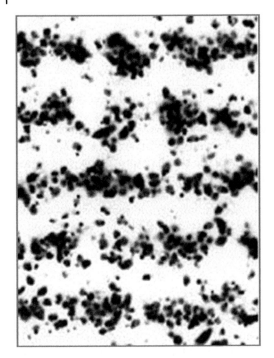

Figure 6.2 Silver halide SEM image. Reproduced by kind permission of Dr T. Rhodes, Harman Technology Ltd.

Nevertheless, the conversion of grains of silver back to translucent silver bromide with a refractive index of 2.23, with the relatively large differential in index from the surrounding gelatin matrix at index 1.50, does offer a formidable potential for high diffraction efficiency if only we can organise the layer in the optimal configuration.

6.2 Pre-treatment of Emulsion

Certain emulsions benefit from pre-treatment, which we might not immediately associate with the use of camera-ready films used in photography or lithography. The purpose of such treatment prior to exposure is primarily to increase the effective emulsion speed, but some emulsions will perform well without any pre-treatment.

Pre-treatment generally can be divided into "wet" and "dry" categories. Because holographic materials are generally extremely slow in photographic terms, it has attracted the attention of experimenters that a chemical or physical method to increase the effective speed of the emulsion would be advantageous. In the case of our own earlier work with pulse lasers, it was very often the case that the maximum power which could be incorporated into a single coherent pulse was often, frustratingly, slightly insufficient to allow optimal exposure of the required recording plate size, leaving the need to "push" the development unsatisfactorily; a means to improve sensitivity was therefore often vital, especially when difficult or demanding conditions of object lighting existed.

Wet methods of pre-treatment include the simplest possible effect, where soaking film simply in water before gently drying it will provide a small increase in speed. This has been explained as possibly due to the resulting dissolution of salts from the layer, which may affect the resulting pAg of the layer. From my emulsion-making experience, there was an observation that emulsion seemed to possess higher photo speed at the "wet-testing" stage, (shown in Chapter 5, Figure 5.5, where emulsion sampled live from the sensitisation vessel was exposed and developed on a glass test plate) than it would appear after completion of the coating procedure to form a solid, dry film.

A controllable and more durable method of pre-sensitisation, which gives excellent results in the case of Colour Holographic BBV materials, is the use of a very dilute solution of triethanolamine $N(C_2H_4OH)_3$ – (TEA), which has been used for many years in the production of *pseudo-colour* reflection holograms due to its ability to produce a controllable and durable expansion of the layer thickness of gelatin emulsions.

6.3 "Pseudo-colour" Holography

Allowing the emulsion to soak in TEA solutions of various concentrations (triethanolamine is a very viscous liquid which sometimes solidifies at room temperature) will result in varying degrees of expansion of the layer. Drying the layer gently, without washing, after soaking in these solutions of up to say 20% w/w in water, for two minutes, results in layer thickness changes which are quite capable of allowing a reflection hologram made by a red laser to replay its image in green or blue light. Prior to the advent of panchromatic (or polychromatic) recording materials for holography, this "pseudo colour" was the only available method to produce multi-colour reflection holograms. Iñaki Beguiristain has become one of the world's leading experts in this technique, and the famous image shown in Figure 6.3 is an example of the perfection of the labour-intensive process to produce a full-colour hologram by chemical manipulation. Iñaki's paper "The Evolution of Pseudo Colour Reflection Holography" [4] gives an insight into the technique including fascinating illustrations of the actual subjects, in comparison with the colour images of the holograms.

It was noticeable to exponents of the pseudo-colour technique with the old Agfa 8E75 film back in the 1980s that the film speed was increased in an attempt to adjust the image colour.

In the case of simple pre-sensitisation, a solution of 1% w/w in water is ideal, as we are not looking to influence hologram colour, but simply utilising the speed increase provided by treatment. Thus, we have a dilute aqueous solution which can be drained adequately with a squeegee or tissue prior to drying very gently in a stream of warm air. This drying must be carried out cautiously in order to avoid drying marks which manifest themselves in the form of striations in the homogeneity of colour in the finished hologram.

This pre-sensitised plate or film can be stored cool for several days before the processing fog level starts to rise seriously ("fog" is the term used in photography for the development of black silver in the absence of exposure to light). We shall discuss the significance of fog development in relation to holography later.

Figure 6.3 Pseudo-colour image of Grommet. Figure courtesy of Iñaki Beguiristain.

6.4 How Does Triethanolamine Treatment Work?

The hypersensitising effect of TEA is, to some extent, a separate issue from its use to swell emulsions for the purpose of creating colour change by fringe spacing adjustment.

Ammonia has been used in the past in the manufacture and hypersensitisation of conventional emulsions, but this was difficult to control. Ammonia has the ability to induce solubility of silver halides. There were multiple explanations for its effect prior to exposure and these include its alkalinity, dissolution of AgHa and the effect on pAg. The "lone pair" of electrons in the vicinity of the nitrogen atom means that this is a polar molecule and, as such, electronic influence may be involved.

The basic structure of TEA is, of course, that of substituted ammonia and we may be witnessing a complex (more delicate) effect of the same type. Reduction of pAg will eventually lead to fog at development, but the effect of the larger polar molecule TEA is perhaps responsible for the better control we achieve by this popular method of pre-sensitising.

Since water itself offers an increase in speed, the observation that emulsions often appear faster in their raw state, before coating, which in my days in emulsion research was sometimes attributed to possible effects such as lack of stress at the gelatin/silver bromide crystal surface, might be relevant. The small advantage in speed, and the larger advantage of longevity of the TEA effect, could, therefore, be a summation of these phenomena; but its advantage for expansion of the layer and the predictable and stable results are invaluable in everyday work in the holography lab.

Other methods of hypersensitisation are available and logically seem appropriate after the previous discussion as they will provide a chemical environment to promote development. Dilute solutions of sorbitol, ascorbic acid, sulphur dioxide, silver nitrate and mercury salts, for example, were mentioned in photographic literature in the past and thus offer possibilities in holography.

As previously mentioned, though, dry processes are also possible. A very low intensity exposure to light has an obvious ability to bring the emulsion to the threshold of development on the H&D curve, but fog must be guarded against, of course, in case excessive exposure is involved.

Hydrogen hypersensitisation of the layer appears, at first sight, to be an attractive option. A relatively non-toxic (albeit highly flammable) reducing agent seems promising. All pre-treatments are extremely inconvenient in terms of the use of films for mass production processes; the idea of taking a roll of film from warehouse storage direct into an exposure device is very attractive indeed; any pre-treatment is naturally inconvenient, but passage through a chamber of gas seems feasible. However, the practical realities of the process are prohibitive. There is a need to de-gas the layer under extreme conditions of low pressure before introducing hydrogen, and the time scales are quite unrealistic.

Obviously, exotic methods such as the above are valuable research tools and certain industrial applications of the holography process may benefit; however, for commercial holography, the simple TEA process is generally more attractive than the alternatives.

I have noted that the application of TEA pre-sensitisation to the Colour Holographic emulsion not only increases the film speed, but provides an additional advantage, in that the holographic image resulting seems actually "improved", as compared to an untreated emulsion simply exposed to the necessary *extra* laser illumination (exposure time), as judged by the ability to provide equivalent exposure, in terms of achieving an appropriate density of developed silver. The TEA sensitised layer appears to be not only faster, but *better*.

Other recording materials may require other pre-treatments. Good guidelines are available from the manufacturers themselves. For some of their emulsions, Slavich recommend pre-hardening treatment, and in some cases, pre- or post-exposure exposure to low levels of white light are advantageous to ensure that we are working on the correct section of the D logE curve discussed in Section 6.8.

6.5   Wetting Emulsion Prior to Development

In view of the fact that we refer to the concept of "volume holograms", we are acknowledging the existence of a finite thickness of the emulsion (typically 6–10 μ in the dry state) in silver halide materials.

To remove residual triethanolamine and indexing fluids, and in order to ensure homogeneous treatment of crystals within the depth of the layer during development, where it could be argued that the surface plunged into warm developer may receive additional treatment as compared to the "distant" emulsion adjacent to the carrier substrate, it might be prudent to allow a short period of soaking in de-mineralised water prior to development. This will then allow developer to penetrate more quickly throughout the wet, pre-swollen layer, with the effect that each section of the volume of the silver halide/gelatin layer is treated more equally.

6.6 Development

In principle, a developer is a mild reducing agent which is capable of distinguishing a silver halide crystal that has been exposed to light from an unexposed crystal, such that it is able to convert the former into a grain of black silver metal without effect upon the latter.

The basic redox mechanism of converting a silver ion into a metal atom is simple:

$$Ag^+ + e^- \rightarrow Ag^0$$

However, the amplification mechanism which is the strength of the photographic process is dependent upon the process mentioned in Chapter 5 whereby a minimum of four photons of incident light is able to create a latent image in a crystal of silver bromide. This latent image represents a physical change to the sensitivity specks on the crystal, which occur during the period of exposure to light and render the exposed crystal developable. In fact, this physical change takes the form of a small group of silver atoms which will act as a catalyst for the reduction of the remaining silver ions in the crystal lattice to form silver metal. Of course, this process is the essence of the amplification which is offered by silver halide photography.

As early as 1908, C.W. Piper [5] wisely observed that "we may look upon development as being of a catalytic nature, the latent image being the catalyst."

Even with our modern analysis capabilities, including electron microscopy, this complex development mechanism still holds an air of mystery.

The latent image can sometimes fade with time, so that "latent image regression" can be responsible for a loss of image density if film is not developed sufficiently quickly. The occurrence of this phenomenon is often dependent upon the manner in which the film was exposed, so that pulse laser exposure has been subject to significant problems in the past with this effect.

Fortunately, in general, the latent image formation process is confined to the individual crystal; in cases where "infectious development" is encouraged, such as in certain graphic arts work, the effect of combining the activity of a plurality of crystals is caused by chemical action of the developer. For holography, with our resolution requirements at an absolute premium, it is vital that agglomeration in general is prohibited at every stage of the recording process, from the dispersion of the crystals in the recording emulsion to the prevention of clumping and infectious effects in both the development and bleaching baths.

So, if we imagine a change which has taken place in the emulsion, crystals which have been subjected to light exposure in the form of an interference pattern where planes of constructive interference manifest themselves as high levels of intensity (shown as cubic crystals with the perimeter highlighted to represent the latent image in Figure 6.4), then the exposed emulsion awaits further activity in the form of chemical action in order to cement this minor physical change into a form where we can actually see the first tangible evidence of the exposure to light.

There are two distinct types of development. These are known as chemical and physical development.

1) **Chemical development**

In this classification, a reducing agent is a component of the developer solution, which directly causes the conversion of the silver ions of the crystal lattice into metallic silver in a process catalysed by the presence of the latent image speck. Without this catalysis, a crystal containing no silver speck does not react to the developer, so

Figure 6.4 Formation of latent image.

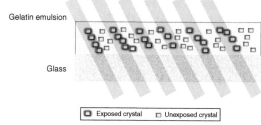

Standing wave of interference

Gelatin emulsion

Glass

☐ Exposed crystal   ☐ Unexposed crystal

that we see that an important property of the developer is that its redox potential is at a level where the silver catalysis at the latent image speck is the arbiter as to whether the reduction of the silver ion proceeds; that is, a relatively *weak* reducing agent.

2) **Physical development**

This is a subtly different process wherein the *silver metal* is supplied by the developing solution itself. Whereas this technique is less commonly used than the chemical route outlined above, it is particularly relevant in the processing of ultra-fine material and deserves some attention as the development method which gave rise to some of the finest progress the world has ever seen in holography, in the form of the famous holograms from the former USSR. There was an exhibition of these holograms called "Treasures of the USSR", which attracted much attention when displayed in London at the Trocadero in 1985.

In this process, the sensitivity speck of silver catalyses a reaction whereby silver metal precipitates from the developer. The process can be carried out before or after fixation (dissolution of the original silver bromide), dictating the morphology of the silver grains produced. In the case of post-fixation treatment, the bulk of the silver halide having been removed, silver is supplied by the developer solution. Without fixation, the influence of silver halide solvents such as thiocyanate or sulphite in a developer ensures that silver ions are readily available from silver halide material which is thus dissolved in situ.

There are hybrid situations where these development categories of reaction can occur simultaneously. Whereas these are highly complex mechanisms, it is quite clear that there is a distinction between them. The presence of silver ions simply does not occur in a fresh chemical developer, but this is not the case when the chemical developer has been extensively used (exhausted). So there is a practical guideline here; for this reason alone, ***never return used developer to the bottle.***

We shall later consider longevity of developers after mixing, but in this case, whereas the ordinary problem of aerial oxidation is a significant problem as regards ageing of developer, it pales into insignificance in comparison with contamination by previously used solutions.

6.7   Filamental and Globular Silver Grains

The silver metal which is produced in the development process can range between discrete globules of crystalline metal and linear filaments. This intermediate distribution of metallic silver is often described as *black silver*. In practice, with the fine-grain emulsions that we deal with in our application, the developed silver frequently appears to have a slight green or red hue, dependent upon the exact configuration of the silver metal: its grain size and shape. In the case of phase contrast holograms, which we seek to produce for Bragg volume holography, the silver metal is an intermediate product which we intend to dissolve selectively for the purpose of "diffusion transfer" in the bleaching process.

Clearly, the very nature of the original developed silver is the most important arbiter of the exact form of the "new" silver halide formed in the "diffusion transfer" bleaching process.

In *The Theory of the Photographic Process*, Mees and James [6] show electron microscope evidence at 50,000 times magnification that the early Lippmann emulsions progress during development from discrete globular crystals of silver bromide to long threads of metallic silver, which fibres appear to grow away from the surface of the crystal, possibly commencing at the sensitivity speck formed by exposure. In larger-grain emulsions, strings of fibrous silver are seen to envelope the original crystal; in the materials we use for holography, as in the case of Lippmann, there may be insufficient bulk in the crystal to ensure its survival as a discrete globular entity.

The process of dissolution of developed silver in the "re-halogenation" step, is an ideal area for investigation by detailed electron-microscope examination, as to how we can develop this process to create orderly planes of high-index silver bromide ($n = 2.23$) in layers that contrast in refractive index with specific planes of gelatin ($n = 1.50$) to create excellent values of Δn.

6.8   The H&D Curve

Hurter and Driffield established a universally accepted system in the late nineteenth century for classifying the performance of black and white silver halide photographic materials, and this important plot, relating developed density with the logarithm of the exposure, is known as the "characteristic curve" or D logE curve (Figure 6.5) for an emulsion. Holographic recording emulsions are, in effect, equivalent to black and white

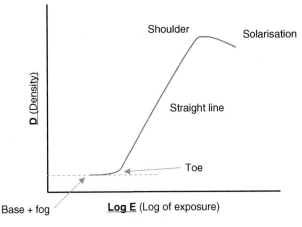

Figure 6.5 The H&D curve (characteristic curve).

photographic emulsions and it is therefore very useful to consider the classical view of what is happening in the recording layer, notwithstanding the fact that we are dealing specifically with performance at the level of microscopic detail; ideally, we would like to make very precise amplitude recordings of sinusoidal density variation at the microscopic (Angstrom) scale.

For any photographic emulsion after development, the characteristic curve represents the conventional measurement of performance. From this, for example, emulsion physicists calculate the quoted "speed" of camera films. Hurter and Driffield regarded the straight line portion as the "correct" area of use of the film. The slope of this portion is the γ (gamma) value of the emulsion.

In a *densitometer*, there is a minimal absorbance when a "clear" zone of developed film is measured, because the base material and emulsion layer have a finite optical density (D_{min}).

At the point where the exposure reaches a level where a latent image is formed, the "toe" is established and, for general photographic purposes, the increase in developed density of 0.10 OD *above fog* defines the "speed" of the film.

Holographic films are rated for their sensitivity in terms of $\mu J/cm^2$. The conventional speed rating of ASA and DIN are of little use in this application – holographic films would rate as a fractional value on the ASA system.

In the straight-line zone of the H&D curve, we have a linear increase in density as the logarithm of exposure is increased. At the shoulder, the film reaches its maximum density (D_{max}), which, for an ultra-fine-grain holography material with incoherent exposure, will normally be of the order of 4.0 OD. Further exposure will then reduce the density in a process called *solarisation*. (This phenomenon is exploited by conventional emulsion chemists in the production of "direct positive" films by chemical enhancement of the principle.)

Optical density (OD) is the logarithm of the reciprocal of the transmittance, and is a really useful measure of the transmission through a filter or developed film. The key values to remember are shown in Table 6.1.

For hologram purposes, we must recognise that, without special microscope facilities, we are interested in recording interference patterns by producing fringe lines and planes at microscopic scale, which comprise a high density of developed silver in the zones of constructive interference and (hopefully) minimal density in the zones of destructive interference by utilising high contrast performance of the film and developer on an

Table 6.1 Optical density (OD) key values.

Optical density	Transmitted light %
0.30	50
0.60	25
1.00	10
2.00	1.0
3.00	0.1
4.00	0.01

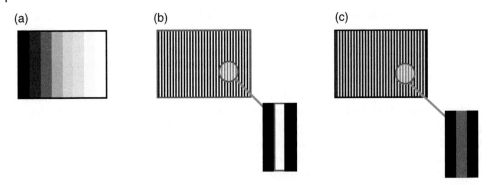

Figure 6.6 Contrast effects. (a) Photographic; (b) satisfactory contrast; (c) reduced contrast.

unusually minute scale. (The resolution of a holographic emulsion such as the historical Agfa 8E75 is rated at about 5000 line pairs per millimetre.)

But, unlike the routine processing of a photographic step wedge represented schematically in Figure 6.6(a) for simple visual or densitometer assessment, without microscopy, we cannot see the evidence of the detail within the recording at the development stage – so we are generally speculating as to its quality, as regards the ability to produce an efficient hologram.

As the development level rises and development fog level increases, although integral measured density increases apparently uniformly, the high-contrast fringe structure shown in Figure 6.6(b) is reduced to an inferior structure (Figure 6.6(c)) whose eventual *diffractive effect* may well be inferior.

In practice, the ideal microstructure comprises a precise recording of the sinusoidal standing wave whose microscopic detail defies, to some extent, the resolution capabilities of the recording emulsion. In fact, we are aspiring to reach a recording quality where each *individual* microscopic fringe cycle resembles the structure represented in Figure 6.6(a).

However, the emulsion has a finite maximum density (D_{max}), in accordance with the H&D curve, so that allowing D_{min} to increase will inevitably reduce the dynamic range of the fringe recording.

For all these reasons it is a good idea during test work to mask a small area of emulsion from the exposure. During processing, this area can then be monitored in safe light for the onset of fog; when such fog appears, we can *assume* that other microscopic areas of low exposure within the fringe structure are also beginning to increase in density, with a deleterious effect on fringe contrast. In conventional photographic developers, compounds such as benzotriazole are frequently included in developer formulations with the intention of reducing fog levels. These would have the effect of extending the useful part of the D LogE curve by delaying the onset of fog during development. They have not been utilised in typical published holography developers.

Another very useful tip, relatively simple to organise in reflection holograms where beams arrive from either side of the film, is to arrange for this masking to allow additional separate zones to witness the object and reference beams, in order to confirm visually the intended beam ratios.

Remember that deleterious fog (undesirable development) can be due not only to excessive development or over-exposure, but also to unsatisfactory "safe lighting".

6.9 Chemical Development Mechanism

With chemical development, we are able to explain a relatively simple regime of processing which accounts for all of the practical observations which relate to the use of the typical "Western" recording materials, such as the Agfa Holotest, Kodak, Colour Holographic and Harman plates.

The developing solution in almost every case will be alkaline, and the pH is generally a guide to the level of activity. Therefore, it is favourable to include a pH buffer component such as sodium carbonate, as well as a stronger alkali, in order to ensure longevity of the original selected pH value. A buffer is regarded as the salt of a weak acid and a strong alkali (or vice versa) and this combination has the effect of regulating the level of H^+ and OH^- ions in solution so as to help with the consistency of processing as the reagents contaminate with age.

Silver solvents such as sodium sulphite and bromide are commonly added to conventional photographic developers in order to enhance development by increasing the level of silver halide in solution. This dissolution feature is effectively on the verge of "physical development" and is questionable in the design of simple holographic developers. However, sulphite is used in the pyrogallol formulation described later as a means of colour control and tanning inhibition.

The basic development action is represented schematically in Figure 6.7.

As previously explained, it is not possible to quote a "best formulation" for a general-use holographic developer, because it is vital to re-optimise for the different recording emulsions and to take into account the recording parameters of the individual holography studio and, indeed, the individual holographic image (such as preferred beam ratio).

Colour Holographics' HRT BBV recording plates have the really excellent quality of extreme latitude in their exposure and processing requirements. Following the necessary pre-sensitisation with TEA, it is possible to achieve good diffraction efficiency from an extraordinarily wide range of exposure levels, beam ratios and processing methods.

Prior to the availability of this material (since the millennium), my own experience was based mainly on the use of the Agfa Holotest and Ilford films and plates. All of these "Western" materials are of grain size specifications which permit the use of simple

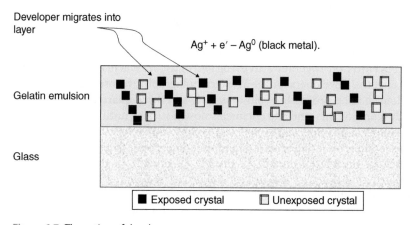

Developer migrates into layer

$Ag^+ + e' - Ag^0$ (black metal).

Gelatin emulsion

Glass

■ Exposed crystal   □ Unexposed crystal

Figure 6.7 The action of developer.

chemical development and reversal bleaching to produce volume phase holograms of reasonable efficiency. With the larger grain size of the Agfa 8E56 (green/blue) and 8E75 (red) material and the older Kodak films, steps must be taken to prevent problems of excessive grain growth if re-halogenating bleach is introduced later.

One of the first highly successful process regimes with the stalwart Agfa 8E75 red-sensitive materials for some of the iconic reflection holograms of the late twentieth century was van Renesse's "Pyrochrome" process, which was discussed in 1981 in a paper by Walter Spierings, "'Pyrochrome' Processing Yields Color-Controlled Results with Silver-Halide Materials" [7]. The developer was based upon pyrogallol, which is actually a relatively powerful reducing agent.

My own first contact with this chemical was its use in the school chemistry laboratory, where "pyro" was mixed with sodium hydroxide to form a powerful reducer to absorb oxygen from air in a eudiometer tube to demonstrate experimentally to students the proportion of oxygen in air. The rapid absorption of oxygen to produce a dark brown solution and thus establish the relative proportions of oxygen and nitrogen in air, shows very clearly the extreme vulnerability of this developer to aerial oxidation.

Pyrogallol was one of the favourite developers of the early photographers, but the disadvantage of its susceptibility to aerial oxidation after mixing into alkaline solution is a clear demonstration that this is a *single-use* bath. In addition to its reducing capability, it takes part in a reaction towards gelatin. This "tanning" effect results in a darkening (staining) of the gelatin layer, but, more importantly, it causes hardening of the gelatin, which, in holographic applications, has been attributed to possible advantages in index modulation, creation of voids, etc., which may improve hologram efficiency.

In the nineteenth century, the effect of hardening the gelatin had been noticed and was used by J.W. Swan (the lighting pioneer) to produce surface-relief effects. Even at that time, the addition of sulphite to the formulation had been recognised as a control handle on stain levels. Both of these effects have been utilised in our holographic application.

The concept of hardened shells of modified gelatin surrounding the individual silver halide crystal whilst it is actively being reduced to silver metal, offers an opportunity for the "reversal bleaching" principle, wherein that transitory silver metal can later be removed to leave a low-density void, *whilst the surrounding hardened gelatin is restricted from accommodating that volume change.* This principle bears similarity to the mechanisms responsible for the remarkable index modulation Δn which offers such advantage in DCG technology.

There is a well-known effect of "super-additivity" for developers generally, which often favours the use of twin developing agents in a solution, so the use of Phenidone or Metol at a low concentration, together with pyrogallol or ascorbic acid as the key developers, provides effective developers for the Western holographic recording materials.

In general, when making processing solutions, use, say, three-quarters of the water volume to mix the components, and add chemicals slowly, stirring until dissolved. Then carefully make the total solution up to the exact volume before bottling. Developers are naturally susceptible to aerial oxidation and, in some cases, to UV light, so it is preferable to enclose the solution in a dark, airtight vessel with no air included. The collapsible concertina bottles are ideal to keep solutions free from air as the volume reduces; the downside is the inability to oversee cleanliness. Some proprietary processing solutions have been made to allow tap water usage, but holography is so sensitive in terms of fringe shrinkage and colour control that even emulsion batch differences are significant, so it is unwise to introduce any further inconsistencies.

6.10   Pyro Developer Formulation

Pyrogallol (1,2,3,tri-hydroxybenzene) was used in early formulations as a lone developing agent. Because the lifetime of the mixed alkaline developer is short, the formulation is mixed in two parts to keep the alkali separate until shortly before use.

In the 1980s, Spierings recommended a simple developer based upon pyro with only sodium carbonate as an alkaline component as the first part of the "Pyrochrome" process credited to van Renesse.

But in his excellent books, *Holograms* [8], and *Practical Holography* [2], Graham Saxby recommended not only the addition of a secondary developing agent in the form of Metol, in addition to a small quantity of bromide, but that a third component, sodium sulphite, was able to offer colour control for reflection volume holograms in the Agfa 8E75 emulsion, and that this solution can also be separated, in order to allow incremental changes in successive iterations of a reflection hologram, and thus provide opportunities for colour control. This component can then be adjusted as an experimental proportion in accordance with the colour of the first test images.

Developer Component A

Pyrogallol	15 g
Metol	5 g
Potassium bromide	2 g

Demineralised water to 1 litre

Alkaline Component B

| Sodium hydroxide | 6.5 g |

Demineralised water to 1 litre

Tanning Control Component C

| Anhydrous sodium sulphite | 100 g |

Demineralised water to 1 litre
Mix the working solution

Immediately before use, equal parts of the developing agent A and alkali B are mixed with stirring. To limit tanning and encourage solubility of silver halide, thus shortening the reconstruction wavelength of the final reflection hologram, one can add up to one equivalent part of control solution C.

Unfortunately, as previously mentioned, the solution has a very limited lifetime after mixing. The solution will darken visibly in the dish and must be disposed of after each use.

6.11   Ascorbic Acid Developers

L-Ascorbic acid is well known as vitamin C, and is, in many ways, an ideal compound, as a non-toxic, mild reducing agent, to utilise as the main developing agent for holography.

The super-additivity effect is utilised, with the addition of Phenidone or Metol as a secondary developing agent.

In mildly alkaline solution, there is little discolouration of the developer or gelatin with this formulation, so a very clear diffractive layer is produced when fine-grain emulsions, such as Colour Holographic BBV plates, treated by this developer later emerge from a bleach.

Despite the apparent propensity for ascorbic acid to hydrolyse in aqueous solution, the ascorbate ion (effectively sodium ascorbate), in alkaline solution of pH < 11, in combination with a small Metol concentration, will keep well in a closed container for many weeks, and will retain consistent activity in the tray during intermittent use for several hours. During periods of disuse, float an identical (clean!) processing dish in the tray to exclude air.

Early ascorbic acid formulations for photography, such as the Agfa developer described by Graham Saxby in *Holograms* [8] contain sodium carbonate with a sulphite content ensuring silver in solution. At Applied Holographics, we eliminated the sulphite solvent and used pH measurement to monitor activity whilst working towards a durable and stable developer for a processing machine where the developer would be held at a moderately elevated temperature (<30 °C) with replenishment at a realistic rate. We added Metol to benefit from the super-additivity effect. A similar formulation is therefore useful for laboratory use, especially because, at lower temperatures than 30 °C, the developer will remain stable and active in the dish for long periods.

With anhydrous sodium carbonate as the main alkaline component providing a measure of buffering of pH, sodium hydroxide is added to increase the activity to a level where relatively short periods of development of the order of two minutes will produce suitable densities for correctly exposed reflection holograms, at 25 °C in the processing room.

6.11.1   Ascorbic Acid Developer Formulation

Ascorbic acid	25 g
4-methylamino-phenol sulphate (Metol)	4 g
Anhydrous sodium carbonate	60 g
Sodium hydroxide	15 g

Demineralised water to 1 litre

Weigh out the two developing agents and dissolve these in demineralised water with magnetic stirring. Then dissolve the alkalis in water. Sodium hydroxide dissolution is exo-thermic. When the solution is cool, *gently* add the two solutions together, with stirring, to restrain fizzing and then bottle with the exclusion of air.

Use the developer to produce densities of ~2.5 OD in holographic plates for reflection work and densities of ~1.2 OD for transmission holograms as starting points for your experimentation for the individual image. Increasing the pH by some degree will increase the activity of the developer, as will raising the temperature slightly.

Remember that although ultra-fine-grain emulsions are apparently capable of very high densities of black silver, in holography we are recording fringe patterns comprising layers, within the depth of the emulsion, of alternating high and low density, so at the coating thicknesses used in holographic materials, we never achieve the densities exceeding 3.00 OD which are sometimes seen in other applications, because it is our intention to develop only something of the order of half the silver halide crystals in the layer; so the D_{max} of the coating to incoherent light is not really relevant in our application. As described, it is very useful to have an area near an edge of the hologram plate which has not been subjected to exposure, protected by the edge of the plate holder or by black masking tape in order to establish the onset of development fog in that zone.

The amplitude modulation of a hologram fringe recording is effectively optimised when the microscopic zones of constructive interference reach their maximum possible density prior to the onset of fog in the unexposed zones, as described in Section 6.8. For example, if an emulsion is massively over-exposed, or if it is left in developer too long or exposed to excess light in the processing room, the density of the intended "clear" zones will rise, as was shown in Figure 6.6(c) and will therefore prevent the film reaching its optimal amplitude modulation. In photographic work, we tend to observe this defect clearly and immediately, but in the holography application it will manifest itself only later, in a lack of diffraction efficiency or signal-to-noise ratio in the final holographic image.

Exhaustion of developer solution occurs on a local basis, and that is the reason why the agitation of the processing tray is always advisable. Developer in the vicinity of a highly exposed zone of the film surface is oxidised by its interaction with the crystals bearing a latent image. By agitation, we can move fresh solution into the area of activity, and ensure that the whole surface of the film is treated equally. Of course, with developer there is also an important "poisoning" effect on the reducing agents in the solution at the surface where aerial oxygen is responsible for the exhaustion of developer, so there is an advantage in organising agitation at a level that avoids excessive aeration of the bulk of the solution – in a processing machine, a floating lid or plastic spheres are often used to prohibit the access of air to the solution.

6.12   "Stop" Bath

When the film layer leaves the developer bath, its emulsion is soaked in active developer. The process of development therefore continues until the developer trapped in the layer is exhausted or neutralised; furthermore, residues may gather in the emulsion layer. An acidic bath is capable of reducing the pH in the layer very rapidly, with the effect that development stops straight away.

It is conventional in photography to use a bath of dilute acetic acid immediately after removal of the film from the developer. Unless we are using very active developer at high temperature for short development periods, which is not generally advisable in holography, there is little advantage in stopping the development suddenly, so a running water bath at room temperature is a good alternative for routine production of masters, etc. This running water dilution of contaminants including spent developer components will have a favourable effect upon the lifespan of the subsequent bleach bath.

In the case where a precise series of experiments with a range of development times is required, use a 2% solution of glacial acetic acid in water to stop development quickly.

6.12.1   "Stop" Bath Formulation

Demineralised water	750 ml
Glacial acetic acid	20 g

Water to produce 1 litre of solution

Glacial (concentrated) acetic acid is a dangerous chemical. Eye protection is absolutely essential when mixing the concentrate, but also the acrid vapour justifies use of a fume cabinet or, at least, a great deal of ventilation. To reduce the frequency of this particular health and safety issue, it is perhaps worthwhile producing an intermediate 10% stock solution for regular re-dilution 4 parts to 1.

Once the development activity is stopped in either of the suggested ways, it is often useful to examine film in the safe light of the laboratory. In some types of holograms, the experienced holographer can begin to amass much information about the exposure and development at this stage.

For example, there is generally a distinct difference in the optimal developed density of silver for transmission holograms and reflection holograms. One excellent way to monitor this in real time is to have available in the dark room several neutral density films attached to a card at densities of 0.6 OD (25% transmittance), 1.0 OD (10% transmittance), 2.0 OD (1% transmittance) and 3.0 OD (0.1% transmittance). This will allow you to compare the developing film in safe light with the filters and achieve consistent diffraction efficiency and noise levels in holograms, day to day.

When making phase-modulated master transmission holograms with re-halogenating bleaches, the density of developed silver is quite critical as a guide to the point at which grain growth in the bleach will lead to scatter and noise problems. The development needs to be stopped at a relatively low density, especially when masters are intended for later use with blue laser light, for example for embossed holography mastering; in this application, *any* excessive grain growth during the subsequent bleaching process will lead to undesirable levels of scatter.

In the developed state, it is possible with image-planed reflection holograms to make very useful assessments of the suitability of the beam ratio used in the shot. For an image-planed hologram, the outline of the subject will be evident in the black silver image; but if this is *excessively* outstanding, you will experience real difficulties with bleaching and scatter levels in the finished hologram. In that case, the amplitude photograph recorded in the film surface will disturb the homogeneity of the surface; this is known as *burn out*.

The stop bath, and in fact any bath in the processing sequence, should not afford any drastic change in temperature from the previous bath, in order to avoid reaction from the softened gelatin in the form of surface reticulation, which will seriously degrade your hologram, especially in the transmission mode where surface relief may be regarded as an inevitable component of the primary image.

6.13 Fixing

Fixation is the term for the process step which is vital in conventional black and white photography, whereby, after developing graduated levels of black silver metal to represent the exposed zones of recording film, the emulsion is "fixed" to remove remaining silver halide crystals. This renders the layer predominantly insensitive to further photolytic activity and thus cements the permanent recording.

In holographic applications, there is a special consideration that we refer to frequently, which does not necessarily apply in conventional photography, whereby the general principle of material leaving the layer inevitably tends to reduce the volume of the layer.

This has generally unsatisfactory consequences with regard to layer shrinkage and hence fringe distortion. In volume holography, the importance of the Bragg condition, discussed in Chapter 2, dictates that fringe shrinkage is generally extremely problematic. For this reason, the use of fixation must always be treated cautiously.

However, in conjunction with certain physical development methods, with emulsions of larger grain size or in combination with re-halogenating bleaches, in cases where shrinkage is not especially deleterious to the required image, fixation and partial fixation, to etch and reduce the size of undesirably large silver grains, is potentially a useful technique.

Proprietary fixers such as Ilford Rapid Fixer are recommended for use in dilute solution mixed 10% in demineralised water. This is based upon ammonium thiosulphate and is rather over-active for ultra-fine-grain emulsions unless highly diluted. Other fixer solutions contain hardening components such as alums and buffer components. Cyanide and ammonium thiocyanate are alternative reagents for the dissolution of silver bromide, but are even more reactive and less desirable for safety reasons, so these are not really suitable for our purpose. The original conventional fixer formulations were based on "hypo" or sodium thiosulphate. For a gentler fixing action for fine-grain holographic emulsions, a simple dilute hypo can be useful without acidification or hardener, and can be used at room temperature.

In the 1990s when embossed holography masters were made with the older materials such as Agfa 8E56 and later the slightly smaller grain-size emulsions by Ilford Ltd, the reduction of scatter in masters for use with the 442 nm helium–cadmium, 458 nm argon and 413 nm krypton lines was a major consideration. Re-halogenating bleaches, of course, offer higher diffraction efficiency, but where signal-to-noise ratio assumed primary importance, the use of amplitude transmission masters was quite feasible, provided an excess of laser power was available for the image transfer of a *less efficient* hologram.

I found that the use of very low exposure levels with object ratios beyond the norm for transmission phase holograms offered a low-density development (<0.6 OD) which could then be fixed for stabilisation purposes. Of course, in a volume hologram this dissolution of unexposed silver halide does involve shrinkage, resulting in fringe distortion, but if a transmission hologram with a reasonable degree of beam symmetry is recorded, then an amplitude hologram in the form of black silver linear fringes is capable of producing a high-contrast, low-noise transfer which does not suffer from the background haze/noise level associated with a bleached phase recording, yet is still able to provide sufficient diffraction efficiency to produce a successful transfer hologram.

6.13.1 Fixer Bath Formulation

Demineralised water	750 ml
Sodium thiosulphate	25 g

Water to produce 1 litre

6.14 Bleaching Solutions

Pre-dating holography, there has been long experience of the need to *bleach* silver halide films, such as the "dot-etching" process which is used to modify half-tone effects in lithographic systems. A strong oxidising agent such as potassium dichromate in acid solution was used in dilute form to allow gradual dissolution of silver metal and such a reaction can reduce the size of a dot in the half-tone process.

We have spoken elsewhere of dichromated gelatin (DCG) as a very effective medium for recording volume holograms, but it should be noted that the present use of dichromate (Cr VI) in acid solution (effectively dilute chromic acid) is really a fundamentally separate use of chromium ions. There are, however, similar hardening (tanning) implications from the by-products.

The essence of *volume holography* is that we are dealing with a "thick" layer of photo emulsion which, after development, contains black silver metal, which will prevent the free passage of light through the layer. The conversion of this black metal to a translucent form of some kind will allow image-reconstruction light to travel through the layer and the bleaching process thus allows us to produce a *phase* hologram from the *amplitude* recording. In the phase hologram, light waves are retarded by index modulation instead of being absorbed by the black metal.

In reflection holography, in the "Pyrochrome" process mentioned above, the use of pyrogallol developer to produce a tanned gelatin layer with black developed silver metal forming a planar fringe microstructure, interspersed contrastingly with unexposed high-index silver bromide crystals (which did not receive a latent image speck during exposure), offers a classical opportunity for an acidic potassium dichromate bleach to operate.

In 1980, Nick Phillips with co-workers A. A. Ward, R. Cullen and D. Porter at Loughborough University, in that era a prominent group in holographic research, reported "Advances in Holographic Bleaches" [9]. A basic dichromate formulation was provided for use with Agfa materials with a relatively pessimistic prognosis for possible diffraction efficiency, which was, to some extent, superseded shortly afterwards by the van Renesse process championed by Walter Spierings.

At Applied Holographics in 1983, we were able to utilise the emulsion shrinkage which results from *solvent bleaching* of this type. In the case of the reflection holograms made by AH, ruby pulse laser exposure at 694 nm into Ilford's Holofilm with ascorbic acid developer fundamentally produced a deep red hologram, whose wavelength meant that the image was subjectively visually unattractive, almost to the point of invisibility, when processed in re-halogenating bleach. However, the controllable shrinkage induced by reversal bleaching with an acidic dichromate solution resulted in a very desirable colour change in the image from deep red to a pleasing yellow-green replay. This "controllable" shrinkage, however, was directly dependent upon exposure level and related to development density, since it was a direct function of the proportion of bleached silver leaving the layer.

Whereas these parameters were directly controllable, the effects of *latent image regression* in early iterations of Holofilm were more difficult to counter. As the latent image regression progressed with time, it was necessary to rewind exposed film rolls before processing, and since the HoloCopier and its chemical processing machine were designed to run at approximately the same speed, each frame was then processed after an equal period of latent image regression. Consistent colour was thus achieved.

This rewinding process on a hand-wound reel-to-reel unit became something of a ritual in Applied Holographics during the company's "maiden voyage" ten-million-unit production run for Nabisco (Figure 6.8). Heroic production technicians Ian Christopher and Bruce Goodman competed for the world record time for the rewind of a 400' roll of

Figure 6.8 "Shreddies" hologram card.

exposed Ilford Holofilm. I was never able to demonstrate any resulting defect in the film so treated, so my own concern at the trail of static sparks lighting up the laboratory during the process went unheeded!

[I use the word "heroic" because Chairman Ossie Boxall, former Datacard MD, had indicated fairly clearly that my own job as production manager was "up for grabs" if the waste rate exceeded 10% – but these dedicated guys achieved a rate of < 8% of waste film!]

In the hologram bleaching process, the dichromate oxidising agent is able to produce silver ions from silver atoms whilst its own hexavalent chromium ions fall from the Cr VI state to the metal's trivalent state, Cr III. There is an element of localised concentration about the formation of these ions, in zones which are, to some extent, image related. The consequence is that zones of hardened gelatin tend to exacerbate the index modulation of the layer, with advantageous effect upon the diffraction efficiency of the grating.

The dissolution of the developed silver metal, tending to result in potential low-density (low refractive index < 1.50) voids, produces a high level of phase contrast when compared to adjacent surviving silver bromide crystals of refractive index ~2.23.

As described, there is also an element of "tanning" associated with use of this bleach because the effect is to produce trivalent chromium. Such trivalent metal ions have the effect of cross-linking gelatin, and are often used in emulsion chemistry as hardeners for gelatin, as we saw in Chapter 5. As regards safety of handling these, it is as well to remember that gelatin is a protein not so very distant chemically from human tissue, and careful protection must be provided to the skin and eyes when handling chemicals, of course. The infamous Camelford disaster resulted from an accidental disposal of tonnes of aluminium sulphate into the town's water supply – simple, apparently "neutral" chemical compounds of this type are not inert!

When such chromium III ions are produced selectively in situ in the photo emulsion in a localised fashion at the site of dissolution of metallic silver from an exposed and developed silver bromide crystal, there is every likelihood that the resulting tanning reaction of the trivalent metal with gelatin *in the vicinity of the silver grain* will contribute to index modulation and diffractive effects. The ionic silver sulphate in the equation is soluble in the solution at this level and is therefore effectively eliminated from the gelatin layer, as shown in Figure 6.9.

$$K_2Cr_2O_7 + 7H_2SO_4 + 6Ag^0 = 3Ag_2SO_4 + Cr_2(SO_4)_3 + 7H_2O + K_2SO_4$$

So, dichromate bleach works well with several developer types and so was the basis of the mass production system produced by Applied Holographics utilising Ilford red-sensitive holographic film, "Holofilm", in the 1980s as well as the van Renesse technique. The disadvantages of dichromate reversal bleach are:

1) Limited index modulation;
2) Toxicity of chromium;
3) Fringe shrinkage;
4) The need to eliminate silver ions from the layer requires the use of demineralised water to avoid silver chloride contamination.

The dissolution of silver in the form of ions in the company of sulphate ions means that anions of chloride (or other insoluble silver compounds) in solution will result in precipitation of silver chloride. For this reason, the dichromate bath components must be dissolved exclusively in demineralised water, and furthermore the wash bath *following* the bleach must also contain demineralised water to prevent the deposition of (light-sensitive) silver chloride on the film surface, and indeed within the gelatin layer.

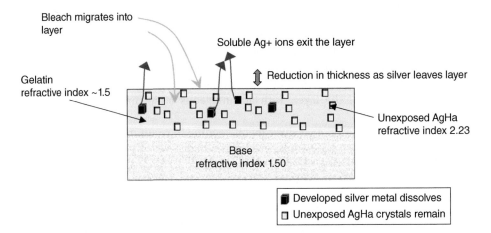

Figure 6.9 Solvent bleaching action.

Once these free silver ions have been eliminated, however, the subsequent completion of the washing process is possible in running tap water.

One definite advantage of the reversal bleaching process is the possible improved signal-to-noise ratio of the diffraction grating. The re-halogenating bleaches which we are about to discuss operate by means of the redistribution of silver halide, and this mobility of material tends to cause both grain growth and increased optical noise and scatter in the layer. However, the straightforward dissolution of silver metal from the matrix of the silver distribution, directly related to the latent image, leaves little scope for loss of fidelity of the microstructure. The reduced potential for high diffraction efficiency with reversal bleaching, induced by the elimination of much of the high-index silver halide, is thus offset by an improvement in the noise level associated with the simplified microstructure.

Of course, a similar bleaching action is quite possible by other laboratory oxidising agents, such as acidified permanganate solution. In this case, a similar bleaching effect occurs; gelatin stain may be removed to some extent, but the manganese VII in the permanganate ion itself, violet in acid solution, is reduced to the lower oxidation states of manganese, which, after drying, have a brown colouration in their own right, but Mn II, for example, does not have any comparable hardening effect on gelatin.

6.14.1 Reversal Bleach Bath Formulation

Demineralised water	800 ml
Potassium dichromate	4 g
Potassium hydrogen sulphate	50 g

Demineralised water to produce 1 litre

6.15 Re-halogenating Bleaches

Re-halogenating bleaches represent a means by which the refractive index modulation (Δn) of the layer can be improved. By this technique, silver atoms are transported within the gelatin layer away from the areas where development has created silver metal concentrations into the areas where zones of the original silver bromide crystals have survived unexposed (without a latent image speck) and are thus undeveloped.

This will allow us to improve the differentiation between the zones which were subject to constructive and destructive interference during the recording process, by producing a phase-modulated microstructure in the layer, with higher levels of refractive index increase than is possible with direct (reversal) bleaching of the developed layer.

The mechanism for this is known as *diffusion transfer*. As we see in Figure 6.10, silver metal which has been produced from crystals bearing a latent image in the development stage can be dissolved by an oxidising agent in the presence of anions which support the solubility of silver cations in water; for example, the sulphate SO_4^{2-} ions that we saw in reversal bleaching.

In the presence of an excess of bromide ions in the solution, these silver ions are quickly re-precipitated as insoluble silver bromide (solubility product 5.4×10^{-13}), in close proximity to their liberation.

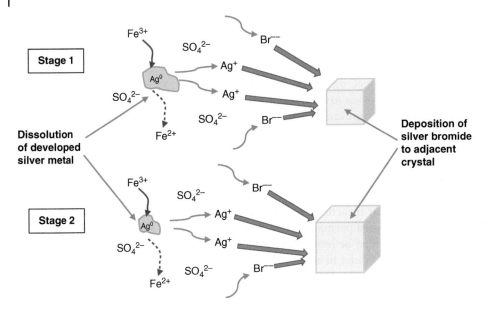

Figure 6.10 The "diffusion transfer" bleach mechanism.

This complex dynamic equilibrium is represented in Figure 6.10.

Silver bromide nuclei will tend to precipitate upon existing local crystals in preference to the formation of new *high-energy* centres of growth. Bjelkhagen describes "diffusion distance" as a measurable quantity which is related to the chemical conditions in the processing bath and the gelatin layer. The conditions to consider in a re-halogenating bleach bath tend to have certain similarities to the crystal growth stage of the emulsion-making process, and there is unlimited opportunity here to experiment with new formulations. We are essentially using methods of crystal growth control to engineer the optimal index modulation of the layer *without* permitting an inordinate increase in grain size.

Large crystals or agglomerations of silver bromide impart high index zones, but are also responsible for scatter. It is often the case that levels of scatter associated with certain grain formations will be relatively satisfactory in the production of holograms to be used in red light, but the challenge of blue reconstruction is far more demanding. There is no reason why complex sequences of controlling growth cannot be influenced by the introduction of other species. In the emulsion-making environment, for example, iodide ions can preferentially deposit upon a bromide surface, with an inherent reduction of solubility and activity.

In the formulation below, it is useful to consider the purpose of each ingredient, so as to allow the reader to experiment with optimising the formulation for their own purposes in accordance with chosen recording materials, hologram wavelengths, beam ratios, etc.

Iron (Fe III) sulphate is a suitably mild oxidising agent with minimal toxicity. The ferric ion in acid solution is reduced to ferrous (Fe II) ion by the absorption of an electron from silver metal:

$$Fe^{3+} + Ag^{0} = Fe^{2+} + Ag^{+}$$

Sodium hydrogen sulphate (sodium bisulphate) is the acid component which is used in preference to sulphuric acid itself, purely for safety reasons. This is an ionic compound which is completely dissociated in water to provide hydrogen ions to the bath. It is thus equivalent to sulphuric acid in many respects and its use was championed by Jeff Blyth, as a leading consultant for hologram processing chemistry, to avoid the serious issue of inexperienced technicians handling concentrated sulphuric acid. Sulphate ions released into the solution join the excess of similar ions from the ferric sulphate.

Silver sulphate is a compound which is far more soluble than the silver halides. Silver ions produced by the oxidation of developed silver metal are thus able to exist in solution.

EDTA (ethylene diamine tetra acetic acid) is sometimes called diamino-ethane-tetra acetic acid, $C_{10}H_{16}N_2O_8$ and is a sequestering agent which has the role of supporting the existence of a (silver) metal ion in solution despite the presence of ions which would otherwise facilitate precipitation.

We are creating an environment for the controlled and selective deposition of dissolved silver and bromide onto the existing surfaces of crystals of silver iodobromide which have survived from the original emulsion, relatively unaffected by the exposure and development processes which have gone before. As discussed in Chapter 5, regarding the growth of silver halide in the emulsion-making process, the balance of desirable crystal growth against the creation of rogue agglomerates of excessive size is difficult to control.

The compound EDTA (Figure 6.11) can be used in the form of its di-sodium salt, ferric sodium mixed salt or ferric salt to introduce it into the ferric re-halogenating bleach solution. Metal ions including silver tend to become entrapped ("chelation") by the multiple negatively charged chain ends of the ionised structure. This is useful to retain metal ions in solution, with restrained freedom for activity, and to control and facilitate the required selective dissolution, transport and deposition of silver in the "diffusion transfer" process.

Potassium bromide (KBr) is the source of the bromide ions which will interact with silver ions released into solution from silver metal oxidised by ferric ions. Silver bromide has a solubility product of only 5.4×10^{-13} in water, whereas silver sulphate is slightly soluble (<1 g per 100 ml). Thus, a solution of ionic silver sulphate *must* precipitate silver bromide when a concentration of bromide ions is present. By achieving a careful balance in these various reagents, it is possible to achieve a controlled deposition of silver, and when ultra-fine-grain emulsions are involved, high diffraction efficiency is possible at relatively low levels of scatter.

$$\left(Ag^+\right)_2 SO_4{}^{2-} + 2K^+Br^- = 2AgBr\downarrow + K_2SO_4$$

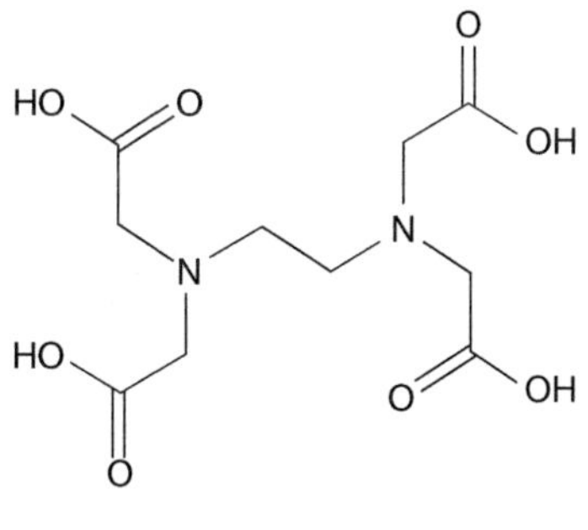

6.15.1 Ferric Re-halogenating Bleach Formulation

Nick Phillips recommended a basic formulation in the 1980s which is suitable for all the finer-grain emulsions and has as low a toxicity as could be expected from any bleach. The addition of sulphuric acid in the original listing has again been substituted

Figure 6.11 EDTA molecule.

by the use of sodium hydrogen sulphate to an equivalent pH value to avoid the undesirable issue of handling sulphuric acid concentrate.

Demineralised water	750 ml
Ferric sulphate $Fe_2(SO_4)_3$	30 g
Sodium hydrogen sulphate	25 g
EDTA disodium salt	30 g
Potassium bromide	30 g

Water to make 1 litre

Add each of the chemicals slowly in the order listed by streaming the powders slowly into water whilst stirring with a magnetic stirrer. Gentle heat in a water bath may be used to encourage dissolution.

Remember that sodium hydrogen sulphate when dissolved in water is equivalent to dilute sulphuric acid solution and is extremely dangerous. Its use does, however, avoid the technician facing the difficulties of the exothermic mixing of concentrated sulphuric acid into water. Safety eye protection should always be worn when using all chemicals and solutions of chemicals.

Make the volume up to 1 litre of solution and bottle it for use. Label clearly and mark the date of preparation. Unlike developers, there is no concern regarding the effects of air at the surface of the bottle or, indeed, the open tray, with the exception of gradual evaporation.

The trivalent iron ion is a useful oxidiser for use in the bleach process as it is not seriously toxic and its pale brown colouration does not stain the emulsion layer very seriously.

However, there are many other oxidising agents which will function similarly in this role. One drawback is that the transition metals are often associated with bright colouration.

The transition between the cupric and cuprous ions of copper also produces a useful oxidation for the purpose of gentle dissolution of metallic silver. This bleach has been used with great success in producing low-noise transmission holograms where the development density is low.

6.15.2 Cupric Re-halogenating Bleach Formulation

Demineralised water	750 ml
Cupric sulphate $CuSO_4.5H_2O$	30 g
Potassium bromide KBr	100 g
Glacial acetic acid	10 g

Water to make up to 1 litre

Stream the chemicals into water in the order listed whilst stirring until all ingredients are dissolved and then make up to 1 litre in volume.

In general, bleaches of the simple formulations shown above are highly stable solutions and the sealed bottle can be expected to have unlimited lifetime. In the tray, the bleaches will last all day – in precise work, it must be remembered that the balance of chemicals, the redox potential and the silver content are changing with every use. These re-halogenating bleaches are undertaking a delicately balanced operation in the phenomenon of diffusion transfer.

For each separate application of the process, we are effectively setting variable target parameters for the process. For example, in the mastering process for embossed holography, we necessarily utilise blue lasers (or even violet lasers). Blue light is especially problematic as regards its propensity to scatter from the smallest particles of matter. Thus, we need to pay special attention to the grain size *not only* of the initial emulsion, which subject attracts much attention, but *growth occurring in the bleaching operation.*

Noise and scatter are inextricably related to exposure level and beam ratio, development density and the properties of the re-halogenating bleach solution. So, for master holograms capable of producing bright, clean photoresist H2 images for embossed holography, it is signal-to-noise ratio which is arguably the most important feature – in the process to shoot a small photoresist H2 there is frequently ample laser power to permit the use of a low-noise master of relatively low D.E. in order to achieve a "clean", crystal clear transfer image.

In the event that a red laser transfer is being used to produce a display hologram, scatter is far less significant. In that case, we might decide to permit increased exposure, shorter beam ratio and higher silver density development, with the consequence that an unusually bright copy may be produced at a quite acceptable level of noise.

6.15.3   Re-halogenating Bleaching in Coarse-grain Emulsions such as "Holotest"

In the case of older emulsions of higher mean grain size of the order of 35 nm, the formulations above, when used to bleach anything more than a minuscule density of silver, will lead to the formation of scatter centres which will render the layer far too diffusing to achieve any level of clarity in a holographic image. It is possible, however, to produce a formulation which includes a reagent to prevent excessive growth, which can form sharp images from red and green laser light, although blue illumination will prove generally to be unsuitable.

The key ingredient of the formulae we used in the past to bleach Agfa film is phenosafranine. (Figure 6.12) This is a dye which has been used in certain photo emulsions in the past as a "de-sensitiser." So this is, in effect, an electron-accepting compound which will trap any photoelectron and prevent latent image formation. It has the capability of adsorption to the crystal surface, just like the dyes which are used in the conventional sensitising role, and it is this property which may be the key to the advantage for the holographic application. There is no reason why other chemicals which will adsorb to the crystal surface may not have similar advantageous properties; in my work at 3M Research, a proprietary colourless electron-accepting compound was developed for similar photographic desensitisation. The diffusion transfer technique in re-halogenating bleaching has significant similarities to the crystal deposition process and inorganic and organic reagents used there might one day unlock a whole area of improvement in control of the growth process.

The dye phenosafranine, which is purple/claret in colour, attaches itself to crystals of silver bromide during the bleaching operation and restrains rampant grain growth whilst usefully desensitising the layer from further photolytic activity.

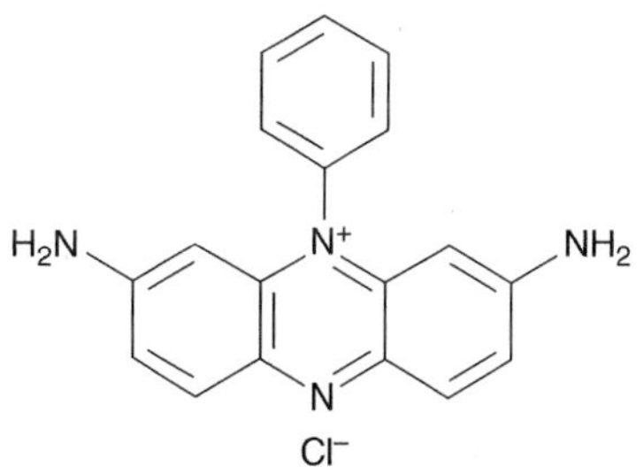

Figure 6.12 Phenosafranine molecule.

This red-dyed appearance of the layer was common during the era of Western display based upon the Agfa Gevaert Holotest products, when much of the work done with helium–neon exposure at 633 nm or the krypton laser red line at 647 nm produced a bright, clean, yellow image in a film dyed deep red. When coated with a black backing, which was often spray paint or PVC laminate, framed display holograms had a very pleasing finished appearance.

6.15.4 Re-halogenating Bleach Formulations for Coarse-grain Recording Materials

In the early 1980s, when the principle of re-halogenation during bleaching began to be understood, the field leaders in hologram processing recognised the value of redistributing material by "diffusion transfer" in terms of building zones of high index within the layer, in conjunction with the added benefit of avoidance of fringe shrinkage.

In the race to achieve such improvements, in an environment where health and safety was significantly less prominent a consideration than it is today, many of the recommendations which were made to the holographer tended to involve toxic chemicals of the type that I would not consider using today. Oxidising agents such as mercuric chloride, PBQ (para-benzoquinone) and bromine were all experimented with and were capable of producing bright holograms in Agfa recording materials.

The Agfa formulation GP-432 was often recommended, for example by Graham Saxby in *Holograms – How to Make and Display Them* [8] and this was based upon PBQ. This compound has the unpleasant characteristic that it takes the form of an incredibly fine powder which diffuses into the air when it is handled. So it must be handled during preparation in a fume cupboard.

In creating wavelength-preserving re-halogenating bleach for reflection master holograms at Applied Holographics, we devised bleach formulations based upon recommendations from Jeff Blyth which enabled us to control quite accurately the reconstruction wavelength of our masters, which were based upon potassium ferricyanide as a principal oxidising agent. Bright red crystals of potassium ferricyanide are a mild irritant, but the greater concern is that, in general, we are considering acid solutions for bleaches. Whereas we preserved the pH at 2.5–3.0 in our buffered solution, the addition of extra acid component is capable of producing hydrogen cyanide gas.

For such reasons of safety, it is therefore wise to direct our attention to the trusted low-toxicity bleach formulae already described. The addition of 0.3% phenosafranine solution in the quantity of 100 ml/litre of bleach can be added to the ferric solution above.

This dye is sparingly soluble and should be dissolved in "wet methanol" as a stock solution. A few millilitres of water are added to methanol and phenosafranine is streamed in; the volume is then made up to 100 ml, with stirring and heating. We are working with a solution close to saturation; when it is added into an aqueous solution, proceed carefully to avoid precipitation.

6.16 Post-process Conditioning Baths

The film or plate emerging from a bleach solution is not always the finished article. The various eventual uses of holograms demand consideration, in that archival stability may be critical in some cases, and yet the finished layer is subjected to a wide range of conditions.

In a finished hologram, we are dealing with a layer of gelatin which contains silver halide, and in some cases, *all* of the original coating weight of AgHa. To achieve archival durability, further photosensitivity must be restricted. Whereas exposure, development and bleaching regimes may have destroyed, modified or otherwise eliminated the original sensitising dyes, the inherent vulnerability of silver bromide to ultra-violet (i.e. sunlight) and blue exposure may remain.

The traditional photographic process of a lengthy wash in running water to finish processing by washing away residual active reagents, perhaps followed by a short final rinse in wetting agent (for example, Photoflo) to aid drying, is not ideal for holographic films and plates for two reasons:

1) The finished layer containing silver bromide crystals is sensitive to "print-out" (darkening) when acted upon by light in the redox neutral condition.
2) Any lengthy washing process will tend to dissolve solutes from the layer with the effect of emulsion shrinkage.

In the case of glass master holograms there is a requirement for longevity. This is especially relevant in the security industry where there may even be contractual requirements for archival storage.

We have mentioned above the effect of phenosafranine which was used to restrain and desensitise the older emulsions at the bleaching stage. But its deep red colour, which was fortuitously suitable for use in reflection holograms made by red laser light, is not suitable for master holograms, especially with green or blue lasers.

The Ilford film product of the Applied Holographics HoloCopier system underwent research into print-out protection. Trials included the inclusion of a final conditioning bath which was tested with 0.5% potassium iodide solution. A dilute iodide bath is able to coat the surface crystal with yellow silver iodide, which is significantly less photosensitive than silver bromide. However, in Chapter 5, we discussed the ability of a minority of iodide ions within a crystal lattice to create defects which encourage sensitivity. At that time, the print-out resistance was slightly improved, but the scatter level and colouration of the layer were regarded as a disadvantage.

In research with Ilford Ltd, we investigated the use of organic materials with some success, including a dilute solution of diquat dibromide. (1,1-ethylene-2,2-bipyridylium dibromide). This material is regarded more commonly as a selective weedkiller and there was concern over its toxicity. There are other desensitising and stabilising materials in classical photographic processing which have been regarded as stabilisers and desensitisers (such as pinakryptol yellow) which could be advantageous in preventing further photolytic action.

But when considering master hologram preservation, there are more simple precautions which are easy to apply.

The ongoing sensitivity of a re-halogenated film is clearly dependent upon favourable conditions for the creation of silver metal. For this reason, the effect of long final washes to neutralise the gelatin layer from the points of view of both redox and pH neutrality is probably an undesirable step.

If we arrange for a final soak in 1% acetic acid immediately before drying, we can create a low-pH layer which does not favour silver formation. Furthermore, a similar "conditioning" bath in a 0.5% acidified solution of potassium dichromate or potassium permanganate will offer protection on the basis of both redox and pH conditions. The dichromate method will leave a slight yellow stain and the permanganate method may cause a pale brown colouration. Either colouration might well prove advantageous as a means of noise reduction when a master is illuminated by blue lasers.

A technique which was used in past times with larger grain size emulsion was to create conditions to lay down colloidal silver in the layer by a solvent development including a silver solvent such as sodium sulphite in a very dilute developer in full white/UV lighting conditions. The layer will then slowly change to a pale red/brown colouration which will tend to reduce blue scatter. For this reason, the method was useful for embossed holography masters made on the Holotest 8E56 plates.

6.17   Silver Halide Sensitised Gelatin (SHSG)

In an era where we have the benefit of ultra-fine-grain silver halide materials as well as the polychromatic photopolymer Bayfol HX, it is easy to forget the astonishment with which the public greeted the sight of the first dichromated gelatin reflection holograms. The realism and brightness of these images set a precedent which could not be matched with Western silver halides.

Researchers such as Professor Nick Phillips reasoned that the advantages of dichromated gelatin were due to the absence of scattering centres such as silver bromide crystals and the index modulation was due to voids exacerbated by the tanning of gelatin to provide shell-like localities capable of protecting against the redistribution of material which might otherwise reduce the index differential between exposed and non-exposed microscopic zones.

Since it was even possible to make successful DCG images in gelatin layers procured simply by the complete fixation of unexposed fresh Holotest plates, it was clear that the advantages of DCG holograms were due not to any special property of the gelatin itself, but to the absence of scatter centres. (When I "dichromated" these pre-coated [hardened] gelatin layers, I was able to make better Denisyuk holograms in the layer than I could with the silver halide mechanism for which the plates were designed.)

DCG work was restricted to the use of the powerful blue and green lines of the argon laser at 488 nm and 514 nm.

Thus, the idea arose to utilise the silver halide simply as a sensitiser to allow use of low-power red, green or blue lasers, before removing it entirely before the viewing stage. The idea was to introduce chemicals into the processing regime which could produce hardened zones of gelatin, in positions dictated by the exposed silver halide. These hardened zones need to be of sufficient durability to allow the complete removal of *all silver* from the layer whilst still effecting modulation of the refractive index, in the same basic way that DCG operates.

For this, Nick Phillips suggested a first development and then re-halogenation to move all of the original developed silver into the unexposed zones. The film was then subjected to further flood illumination and then the second development to convert all of the silver halide to silver metal. An acid dichromate tanning bleach will harden gelatin zones whilst removing the silver entirely. After washing the silver away, the film is plunged into solvents à la DCG and a phase image results.

Other experimenters have used quite different sequences to achieve index modulation in the finished layer without the presence of silver and have reported good results. There is almost no limit to the various ways in which one could handle film to encourage various methods of tanning and hardening in a configuration related to the microstructure of the standing wave.

The advantages of such a system, if commercialised, would be:

1) *All* process silver could be recovered.
2) Silver halide sensitivity levels to all visible laser wavelengths.

The questions remaining to be answered are:

1) Can the problem of DCG susceptibility to atmospheric degradation be solved?
2) Is it feasible to produce predictable and controllable image colour?

6.18 Surface-relief Effects by Etching Bleaches

In order to allow relief masters to be produced using silver halide (and therefore any visible wavelength laser), it is possible to use a hybrid of the etching bleaches originally published by Kodak.

These bleach formulations were designed to enable Kodak lithographic (precision line) plates to produce a "direct-positive" image. The formulation is invaluable as a method to produce high-contrast artwork plates for 2D/3D holography, as required for foreground block-out masks, described in Chapter 9, which comprise islands of dense black silver on a plate otherwise devoid not only of silver halide but also of gelatin!

After drying and hardening, these plates provide artwork with completely non-scattering areas of clear glass, which can be cleaned after use by gentle polishing. When used as artwork to incorporate planar graphics into a hologram, these can be used with index-matching agents where necessary, as a stack of "2D/3D" artwork planes, with the ability to carefully clean them afterwards for further use, since no gelatin matrix exists in the clear areas of the plate.

6.18.1 Kodak EB4 Formulation

Part A

Distilled water	600 ml
Copper II chloride	10 g
Anhydrous citric acid	150 g
Urea	150 g

Water to 1 litre of solution

Part B
3% aqueous solution of hydrogen peroxide

Parts A and B are mixed in equal volumes immediately prior to use.

Essentially, the Kodak lithographic plate, or Agfa Millimask or 8E56 Holotest plate is exposed to a line image mask by a contact process with white light exposure.

Kodalith developer is mixed 1:3 with water and then used to develop the plate to as high a density as possible whilst avoiding the onset of fog. Other developers can also be used.

Use a 2% acetic acid stop bath to prevent fogging.

The plate is then placed into EB4 at room temperature and agitated. The room lights are turned on to allow close examination of the plate as swirls of silver metal and gelatin dissolve into the bleach. A soft brush is used to gently encourage reluctant detailed zones of silver to dissolve.

It is possible to see the image detail in the surface of the plate in the form of the slightly opalescent gelatin layer in contrast with the crystal clear zones of stripped glass, and the technician quickly gains expertise in the swabbing or brushing of the evacuated areas.

The plate is then washed in water before returning to the developer bath in full lighting in order to develop the silver halide in the remaining areas of gelatin. After maximum density is reached, the plate is washed in running water.

A bath of hardener can be applied to return the gelatin to a state of improved durability after its rather abusive treatment in the urea/hydrogen peroxide containing bleach.

Fixing solution with hardener is suitable for this purpose. Specialist solutions containing formaldehyde or inorganic aluminium sulphate, alum or chrome alum can alternatively be used.

After rinsing briefly in distilled water, the plate is dried very carefully in warm air.

The principle of operation of this bleach is that the attack on developed silver by copper ions effectively catalyses adjacent attached gelatin, softened by the presence of urea and hydrogen peroxide, to *denature*, with the effect that it dissolves in the bleaching solution. This dissolution is delicately balanced with regard to the gelatin type and hardness. It is necessary to modify this formulation to suit various emulsions and certain highly cross-linked coatings are very difficult to accommodate.

However, the complete denaturation required for the reversal effect described does not apply to the principle of using these basic ingredients to exacerbate surface reticulation. With a similar bleach, using reduced urea and increased peroxide, direct from ascorbic development, I have produced holograms with strong surface-relief gratings, indicating that impressive resolution with regard to gelatin dissolution is possible from this general principle of operation.

6.19   Photoresist Development Technique

The acquisition of a bottle of Microposit 303A developer to process your exposed photoresist plate can be the beginning of a torturous experience if the intention is to produce state-of-the-art resists for the creation of commercially acceptable nickel embossing shims.

The word "developer" as applied to photoresist is a completely different technology to the "development" of silver halide. The action of exposure to violet or blue light in a positive photoresist is to render the affected zone subject to solubility in alkaline solution.

The Microposit developer is a strong solution ($\sim$1.7 N) of caustic alkali in water with significant addition of surfactants, which make it possible for the diluted solution to dissolve exposed material out of the layer and to remove it without deposition on other parts of the surface.

Various exponents use quite different dilutions of the stock solution at room temperature, or at elevated temperature. The standard recording material is the Shipley S1800 resist, which is a pale-straw-coloured layer, but which is most often coated upon 5 × 4" or 8 × 10" glass substrate, which is frequently coated with an opaque deep brown iron oxide under layer, that acts as an anti-halation device by absorbing laser light which would otherwise reflect back into the recording layer from the rear surface of the glass. (The recording of surface-relief transmission holograms for embossed holography dictates that the laser beams arrive from only one side of the plate.)

For this process, there is a need for copious quantities of distilled water, and a means of driving surface water from the surface of the plate, which may be, for example, a pump to supply a jet of filtered compressed air, a cylinder of pure nitrogen gas or air or even a very powerful hand-held hairdryer type of device, which is able to provide a very powerful directional stream of cool air.

The process has resisted all attempts for automation in my experience. There is certainly a case that visual inspection of the plate, with an ability to adjust the development time in accordance with the appearance of the image, is a vital advantage.

The surface-relief hologram is naturally especially sensitive to any spurious contamination of the surface whatsoever. There is a real difficulty in ensuring that the resist that is dissolved from the linear microstructure forming in the plate surface is able to disperse entirely into the developer solution without random deposition on the planar surface, with the effect of producing white haze deleterious to the subsequent production of a master shim.

The development action is a dissolution of the coated material, and there is a differential in the potential for exposed and unexposed microscopic zones to dissolve. The coating is very thin (1.5 µ) and it is very easy to reduce the thickness beyond optimum, so it is absolutely vital to create a definite processing routine with time and temperature control, in conjunction with control over the beam ratio and exposure energy to ensure consistency.

The processing operation is carried out in yellow fluorescent light; there is no advantage in working at lower levels of illumination, because resist is sensitive only to blue and violet light.

To the uninitiated observer, the process of manual development of an 8 × 10" resist plate is a remarkable spectacle to the point of hilarious entertainment. In order to ensure the dispersion of solute away from the plate surface, it is necessary to plunge the plate face down into a *deep* tray of temperature-controlled 303A solution mixed, say, six parts distilled water to one part of concentrate. The plate must be rapidly (violently!) agitated under the surface of the solution, of course with no contact with the base or sides of the large tray. The correct fringe profile is dependent upon this agitation, since the developer solution is rapidly poisoned, not only by aerial action at the interface (absorbance of carbon dioxide) but also by the action of dissolved resist.

So this process is one where extreme attention must be paid to safety issues. Any splashing from the tray involves caustic alkali, so that glass screens, goggles, gloves and full protective overalls *must* be used. At the end of the timed development of, say, ten seconds, the plate is removed rapidly from the now contaminated solution and plunged rapidly, with agitation, into a tray of distilled water. It is then flipped face up to examine the image.

This is now an opportunity for a purely visual assessment under yellow light. The technician must decide whether there would be any advantage in returning the plate to the developer for one or two seconds. If so, the whole cycle is repeated.

When this is completed, the image can be carefully examined for defects under a spotlight. This should be a permanent fixed light that can be used to compare each resist emerging from the lab. That way, you always have the ability to keep a successful plate at hand to make a subjective relative assessment of your new work. The subjective view of holography is an underestimated quality parameter!

Defects which can be recognised at this stage under spotlight are:

- Concentric phase rings due to overdevelopment or over-exposure;
- Surface imprint of the image detail too strong;
- White scattering deposit on the surface;
- Insufficient brightness of the image.

When you are truly satisfied with your plate, it moves forward to the next stage of metallisation. Resists can be metallised with a conductive coating by either electrostatic vapour deposition in a vacuum chamber or, alternatively, by chemical spray deposition.

I have seen both methods produce magnificent results, and in the case of the spray technique, experience of actually watching in real time the sudden amplification of the image brightness, as shiny silver metal lines the microstructure is truly unforgettable. Be aware that the infinitesimally thin layer of silver metal has entered into the "valleys" of the surface-relief microstructure, so as to allow us to witness diffraction from the resulting relief microstructure of the reflective metal coating the plate surface, effectively the "reverse side" of the coating.

The resulting silver-coated resist plate is used as an electrode in a nickel sulphamate plating bath to grow a first, or "mother" shim. The metal is electrolytically grown to a thickness where the nickel layer can be separated from the resist-coated glass at one edge and "pulled" from the substrate. Before beginning an embossing job, sufficient shims will be grown from the original to allow replacement of worn shims or duplication of production machines. Successive generations of shims will provide "wrong-reading" and "right-reading" as regards graphics whereas the phase polarity of the fringe structure is unimportant as regards diffractive effect.

Notes

1 Bjelkhagen, H.I. (1995) *Silver-Halide Recording Materials: for Holography and Their Processing*. Springer (ISBN: 978-3540586197).

2 Saxby, G. (2003) *Practical Holography*, 3rd edition. CRC Press (ISBN: 978-0750309127).

3 Electron Photomicrograph by Harman Technologies Ltd, Ilford Way, Mobberley, Knutsford, Cheshire, WA16 7JL.

4 Hologram by Iñaki Beguiristain (with permission from Aardman Animations Ltd) www.inaki.co.uk and "The Evolution of Pseudo Colour Reflection Holography" for the 7th International Symposium on Display Holography at St Asaph in 2006.

5 BJP 1908: The Nature of Photographic Images.

6 Mees, C.E.K. and James, T.H. (1966) *The Theory of the Photographic Process*, 3rd edition. Macmillan (ISBN: 978-0023800405).

7 Spierings, W. (1981) 'Pyrochrome' Processing Yields Color-Controlled Results with Silver-Halide Materials, *Holosphere*, 10(7 and 8), 1.

8 Saxby, G. (1980) *Holograms: How to make and display them*. Focal Press (ISBN: 0-240-51054-2).

9 Phillips, N., Ward, A.A., Cullen, R. and Porter, D. (1980) Advances in Holographic Bleaches,*Journal of Photographic Science and Engineering*, 24(2), 120.

7

Infrastructure of a Holography Studio and its Principal Components

7.1 Setting Up a Studio

It goes almost without saying that the reputation which quite correctly accompanies holographic optical systems regarding vibration and positional stability almost inevitably suggests accommodation on a ground floor, except in exceptional architectural circumstances.

Fortunately, the logistical problems of providing three-phase electrical supply for lasers as well as water cooling systems, in the form of externally situated chillers, have been alleviated recently by the advent of solid-state laser systems, which are rapidly replacing the argon and krypton gas lasers, whose inherent inefficiency formerly led to significant difficulties in absorbing the massive quantities of "wasted" electrical energy into cooling water. That cooling water was responsible not only for difficulties in avoiding vibration in the laser head, but, of course, required the expense of further electrical energy, in chillers, to deal with the ecological difficulty of producing large volumes of warm water. Further, the need to control very tightly the input temperature of (demineralised) water into the laser head involved expensive precision heat exchanging systems to interface the laser with its external chiller.

By definition, we are recording holographic interference patterns whose principal structural dimension is of the order of one quarter of a micron. This means that any environmental motion which results in laser beams involved in the creation of a standing wave (as described in Chapter 2) moving by, for example, one tenth of a micron, will result in almost total loss of the diffractive microstructure recording.

Some of the installations I have worked with have, in fact, been positioned in basements, which may appear at first sight to be ideal for structural stability. However, there are still issues of access and noise from above, so it is extremely difficult to define the perfect environment for holography origination studios.

To help to decide upon a suitable site for a holography system, we should look at the criteria which influence success. During years of involvement with holography mastering systems, we have seen the technology for table suspension develop rapidly. Early installations typically comprised concrete or granite tables suspended upon car tyres containing inflated inner tubes, or upon steel sprung cushion feet; but later on, techniques involved lighter proprietary honeycomb tables with pneumatic piston assemblies, and finally ultrasonic/piezo-type electronic levelling systems. Figure 7.1 shows a pneumatic breadboard table with a silent pump below.

The Hologram: Principles and Techniques, First Edition. Martin J. Richardson and John D. Wiltshire.
© 2018 John Wiley and Sons Ltd. Published 2018 by John Wiley & Sons Ltd.
Companion website: www.wiley.com/go/richardson/holograms

Figure 7.1 Pneumatic suspension of honeycomb table.

7.2   Ground Vibration

Clearly, the function of the optical table suspension system is to prevent transmission of ground-borne vibration through to the optical systems. Dependent upon frequency, the various types of table support legs will reduce or eliminate these problems, but it is no coincidence that two of the largest commercial holography companies in Britain, De la Rue Holographics and Applied Holographics (now Opsec Ltd), originally from city locations, initially chose to set up their operations in very similar environments in the grounds of countryside mansions, in order to avoid obvious heavy vibration from roads, railways and heavy industry. Applied Holographics' environment in Braxted Park, Essex, where I worked for 14 years, was additionally protected by its surrounding lake from vibration from the A12 trunk road. However, it is also true to say that one of the original heavyweight granite tables on which I made my first serious holograms eventually cracked the floor of the aged building on its gradual route towards that lake!

Having successfully dealt with ground-borne vibration, however, holography shows no mercy and both companies became rapidly aware of the effect of flight paths; there is nothing like a low-flying Chinook helicopter to provide a disturbing very low-frequency shudder to the surrounding countryside. It is also no coincidence, either, that both companies later relocated into urban areas as the technology developed; and the fact is that simple access for staff and visitors is an important factor in a truly commercial environment.

Specifically, the approach to building a studio may be influenced by a range of factors which dictate less than a free choice of location based upon environmental advantage. The ability to design a purpose-made building with the correct isolation is generally more important than geographical location nowadays, but, of course, finance remains the dominant feature.

After moving away from their original rural location, a laboratory built upon a raft foundation isolated by pliable under-floor vibration insulation enabled the studios in DLRH to perform with excellent stability, although close to busy main roads. Applied/Opsec also left their original isolated site with a successful result.

Doors which are attached to the building structure are often a major source of vibration, which seems to transmit through the floor and table suspensions. The details of the attachment of the doors to the resonant structure of the building is vital, and in one particular location I experienced a situation where a relatively distant door caused major stability difficulties to the optical table whereas other, much closer doors were almost inert as regards movement problems.

However, conversely, I have also experienced several times the ironic situation where the failure to switch on the table suspension system went unnoticed until after successful holograms had been made! My former colleague, Paul Dunn, reported successful holography despite nearby builders using pile drivers during exposures. Such apparent irregularities do nothing to diminish the undesirable "black art" reputation which holography tended to attract in its early days.

It is useful, with pneumatic systems, to be able to do major setting-up work on the table whilst the system is inactive, as it is so disconcerting to have the table move whilst working, in the event of leaning upon the surface.

Electrical supplies to equipment on the table need special consideration. Cables draping from the table can represent a major tripping hazard in a safe-lit studio, and may even bridge the isolation system to some extent. Sockets which can be arranged in the ceiling above the table or on panels supported between the isolation legs, as shown in Figure 7.1, are invaluable for these reasons.

7.3 Air Movement

Of course, the "movement" problems I have repeatedly experienced, associated with doors closing, as mentioned previously, could also be associated with air movement; a shock of pressurised air can feasibly travel through a building.

One critical feature of modern building is the apparent ubiquitous presence of air conditioning and central heating systems, which are, in general, anathema to successful holography. It is often difficult or impossible, in view of building regulations and modern building practice, to eliminate these systems without significant administrative and practical difficulty. It is, nevertheless, virtually impossible for holographic systems to function whilst air flow and temperature changes are occurring in the way that they do in a typical air-conditioned building. The building regulations for air change and window space etc. must then be circumnavigated, sometimes at significant cost, in order to allow a studio to function successfully. In essence, the ability to change the air in a studio sufficiently to keep a fresh atmosphere is at odds with the need to avoid air movement during exposure, but as long as we have the ability to switch air circulation "on" and "off", then we can achieve both aims.

One of the most radical solutions I have seen to air movement difficulties was the introduction of a complete double glazing enclosure upon the table. Notwithstanding the significant burden of weight upon the pneumatic suspension system, the enclosure provided the opportunity for long exposures without fringe lockers and without motion problems. One of the remarkable visual features of such an enclosed system is that, after a period of time with the windows closed, there is a spectacular reduction in the visibility of the laser beams; the settlement of dust from the air occurs quite quickly, and it is then difficult to see the laser beams inside the unit. Such an enclosure is an excellent advantage where feasible. But many holographers have improved their air stability in the past with a cover fixed above the table in order to support flexible curtains. In this case, plastic curtains are more suitable if dust issues are to be avoided.

Air movement on a smaller scale is also something which must be monitored in the studio; streams of warm air from control computers and laser heat sink fans can produce local movement of air which must be prevented from influencing the passage of laser beams around the table. For this reason, the lasers can usefully be confined to a corner of the optical table which is shielded or insulated to confine such unavoidable movement of air to areas where less harm is done to the stability of the laser paths.

7.4   Local Temperature Change

By far the best solution for enabling a studio to function properly is to utilise exclusively natural ambient temperature wherever possible. We have found that this is relatively easy in temperate climates but more difficult in parts of the world where cold weather dominates. As soon as localised heating is brought into play, we run the risk of introducing temperature gradients which will be deleterious to hologram recording. Even when the laboratory is heated only during hours of disuse, principally at night, the heavy table will store that heat and will then produce rising currents over the table when the air itself cools in the room; equilibrium is what we are really searching for.

This method of allowing nature to dictate the working conditions does not always make for a comfortable environment for the holographer, but such inconvenience is more than justified by successful results; the frustration of producing defective holograms afflicted by black shadows is considerable, but is avoidable.

Microcosms of the temperature gradient problem occur when small components such as electromagnetic shutters remain powered whilst the shutter is open. More sophisticated shutters are "bi-stable", in that they settle, without power, at both the open and closed positions. A coil kept in a powered state during the open period can introduce convection currents capable of causing air movement and density (refractive index) change sufficient to disturb the continuity of a laser beam passing through.

A separate example of unsatisfactory temperature variation in the holography lab is seen when recording materials, wisely stored in a cool environment (refrigeration) to extend their lifetime, are brought into a warmer studio. Condensation can have serious moisturising effects on the emulsion as well as thermal expansion problems and can result in movement – plates need to be in the studio to acclimatise for some hours prior to use. The same principle applies to silver halide plates which have been subjected to chemical pre-treatment and then dried in warm air – allow them to equilibrate completely with the studio environment before use.

7.5   Safe Lighting

Suitable lighting in a holography laboratory is often difficult to specify. So often the idea arises to paint the studio walls and ceilings black with the idea of avoiding reflections off the surfaces, but my own preferred approach is almost the opposite: white surfaces surrounding the work area allow you to silhouette items requiring attention to detail whilst keeping the general level of lighting much lower.

The need to enter and exit the room whilst recording material is exposed on the table, which, unlike photographic lab work, is a frequent requirement in holography, requires that "light traps" should be arranged at the entrance of studios. Ideally, this is a double-door arrangement so that the inner remains closed whilst the outer door opens. In this case, by all means blacken the walls and ceiling to avoid flashes of external light when the double-door system is abused; but if you paint doors black inside the trap, experience shows that it is as well to leave a luminous marker on the surface at eye height!

The trap is safe lit, so it is then possible to enter the lab without compromising the safety of the plate. This trap can also be associated with laser interlocking; one of the issues associated with the use of Class 4 lasers is the avoidance of any opportunity for laser light to impinge upon unsuspecting or untrained people entering the studio, with or without permission.

One of the most frequent mistakes made in the dark room is to over-react to the need to keep the light levels low. It is foolish to work in light levels which are orders of magnitude lower than is necessary, because precision work of the type carried out in a holography studio requires as much visual assistance as is realistically possible *provided the recording material is not influenced.*

In fact, most of the older generation of photographic safe light filters are excessively absorbent for the purpose of protecting holography materials.

First, it is necessary to assess the use to which the laboratory is to be put. There are two aspects to this:

1) The likely recording materials to be used.
2) The approximate likely time span for leaving these materials out on the tables with exposure to the ambient lighting – a situation which is not common in photographic work. The use of "step and repeat" systems might involve the photosensitive plates being exposed to ambient subdued light for literally hours on end, and this must be accommodated. I have seen actual evidence of such plates reducing in efficiency with every successive exposure due to gradual pre-exposure "fogging".

The first category of the above is really quite easy to break down into the appropriate groups. For example:

a) Silver halide materials of sensitivity of the order of 100 μJ/cm^2;
b) Photopolymers of multi-chromatic sensitivity of the order of 10 mJ/cm^2;
c) Photoresists of monochromatic blue sensitivity of the order of 20 mJ/cm^2;
d) Dichromated gelatin of lower sensitivity.

Clearly, the spectral sensitivity curve for each material is a very useful guideline and can guide us towards the selection of the most appropriate colour filter or lamp. A green-sensitive silver halide film is clearly best handled in red safe light, whereas it *appears* that a "panchromatic" material is best handled at the wavelength which is at minimal absorbance in the spectral sensitivity plot.

In practice though, for panchromatic (or multi-chromatic) materials, the sensitivity of the human eye is a very important factor here, because the highest sensitivity of *scotopic* (dark-adapted) *vision* is in the 500 nm area, so in the case where there is no really advantageous zone of insensitivity due to a flat spectral sensitivity curve, photographic experience has taught us that it is appropriate to allow the eye's most efficient wavelength (green) to be used – in that way, total light levels can be reduced whilst the technician's eyes are most advantaged. In some cases, the technician's preference for lighting colour might be the arbiter; some people find low lighting in amber less claustrophobic or nauseous than strong red or green filters.

The blue-sensitised Shipley resist materials (nominally for the ~436 nm mercury peak) have very low sensitivity to other wavelengths, so bright yellow fluorescent lighting can be used for this material without deleterious effect. In fact, a room for processing photoresist realistically *must necessarily* be lit in this way, because the demands for the wet processing technique are a highly visual procedure; to work in unnecessarily subdued lighting would almost certainly prevent the complete optimisation of the holographic image and the elimination of frustrating surface defects (see Chapter 6).

Photopolymer materials which are sensitised to colour such as the Bayfol HX films have surprisingly high red and green sensitivity, and perhaps the best option is to use amber light at the appropriate level. One very useful way to achieve this is by the use of dimmer switches with tungsten lamps, coloured or covered by filters in orange gel, such as those by Lee Filters [1].

The second aspect (point 2) in the list above, refers to the realistic assessment of the likely time period for which film will be exposed in the laboratory. Holographic recording materials, unlike conventional photographic films, tend to be left open on the optical table to ambient lighting for long periods. In the special case of step-and-repeat systems, and stereogram origination methods, the recording plate may, in some cases, be exposed to the ambient lighting for several hours, as mentioned; but all holographic systems tend to result in unusually long exposure to safe light, in comparison with typical photographic practice.

This leads to an obvious experimental method to make a practical assessment with the recording material which is to be used in the laboratory. That is, to cut strips of the film to be used routinely in your studio and to leave them uncovered for appropriate periods of time in order to establish practical guidelines for the most suitable level of light.

Thus, a silver halide film or plate carefully covered on part of its surface by a tape mask and left in the workspace while successive strips of the black tape are removed during a period of, say, one hour, to leave a range of exposure times, can be developed in the recommended developer and then stopped in dilute acetic acid. If the areas exposed to the safe light show a significant density above the level to which the permanently covered zone develops, then the "safe lighting" is not "safe" – it is too bright. This is a simple practical assessment which is suitable for silver halide work.

The use of materials such as photopolymer and DCG make safe light selection a little more difficult, since it is more complicated to establish such a simple test of the damage caused by safe light. But when lengthy periods of research or production are to be undertaken in a laboratory, it is worth significant levels of effort to establish that the process is not being compromised by the ambient lighting. For example, photopolymer film can be exposed as described above for realistic variable periods to safe light before the film is used to make a hologram. If the trial safe light exposures are deleterious to the quality of the hologram, then further investigation is required. As mentioned in Chapter 5, remember, it is *not impossible* for pre-exposure of film to improve results (pre-latensification).

7.6   Organising Your Chemistry Laboratory

The processing room has quite different requirements to the optical studio and it is difficult to justify these functions sharing the same room.

I believe it is good discipline to acknowledge the completion of the exposure process, by enclosing an exposed plate or film in a light-tight safe box (generally a well-made box of the type in which recording plates are supplied) for transport to a separate processing area. The optical studio contains delicate and valuable optical components and it is difficult to justify wet processing in the same area – relative humidity is an important factor in the manufacture of colour-controlled master holograms. Unlike the optical studio, suitable ventilation and airflow are required and ideally laminar flow enclosures for chemical mixing.

The chemistry processing room for silver halide work should, ideally, be equipped with a "wet-table" where processing trays are able to stand in temperature-controlled warm water to ensure temperature consistency. Temperature probes, pH meters, top pan balances, magnetic stirring facilities, densitometers and drying cabinets are all valuable assets for high-quality work. A conventional dark room clock with illumination that does not compromise the safe lighting is a distinct advantage for consistent results. In the halcyon days of photographic home-processing, it was possible to acquire a wall clock with remote control starter, red LED lighting and loud seconds countdown acoustics, which are surely a formality in the "smartphone" era!

Remember that although silver halide, with its aqueous processing, is relatively forgiving towards dust contamination apart from the critical period where the plate is squeegeed and dried after pre-treatment or post processing, this is certainly not the case with photoresist, where a dust-free environment and clothing are essential for quality results.

Forced air filtration is really necessary for resist processing areas, and copious quantities of demineralised water are required at high rates of delivery. Plates are dried by high-pressure jetting of clean nitrogen or air to remove surface water; and it is to be remembered that nitrogen build-up in the room is to be avoided – here, air change and laminar flow cabinets are certainly required to ensure that technicians are not subjected to unsatisfactory nitrogen proportions in an enclosed processing area. Nitrogen is not poisonous, of course, but neither does it support life!

In the design of processing rooms, the most reassuring and convenient of features is a floor drain. It is inevitable that spillage will occur, especially in the processing of photoresist. It is also likely that, from time to time in silver halide work, a wet-table drain will be blocked and the ability of a technician to squeegee and mop a quarry-tiled floor after such an event into a convenient floor drain will ensure the longevity of a clean, healthy and safe environment, whilst, most importantly, preventing any disastrous flooding.

7.7   The Optical Table: Setting Up the Vital Components

The general heading of routinely organising a working optical table for holography involves a number of operations which will rapidly become simple routines with accumulated dexterity and experience. These include operations such as setting up collimating mirrors and lenses in precise alignment, and the regular process of spatial filtration of laser beams.

Many of the various optical components of an optical table are delicate and susceptible to mechanical damage or overheating by Class 4 laser beams. Surfaces are susceptible to dust contamination, which may rapidly be fused into the surface by pulse lasers or high-power continuous wave (c. w.) lasers; but keeping the environment free of dust is certainly not aided by the incompatibility of the holography process with air-filtration systems used typically to achieve clean room conditions.

7.8 Spatial Filtration

We have experienced the frustration of many lab technicians with the process of setting spatial filters in operation. The ability to tune a spatial filter from scratch in a matter of seconds has become something of a speciality which is sometimes seen as a basic test of the holographer's skill, and has even been used in the practical job interview process.

An appreciation of the mode of operation of a typical filter will make the practical problem much less daunting. There are many different styles of spatial filter on offer from the various manufacturers, but, in essence, the principle of operation is simple.

We are dealing with powerful laser light sources with beams focussed to high concentration; even diffuse reflections can be dangerous and the technician must use suitable eye protection. This eye protection, however, whilst attenuating laser light to a safe level, must allow the technician a certain view of the laser beams at a suitable intensity when required. Health and safety executives who are not familiar with holography will often tend to insist that technicians simply wear safety glasses opaque to the laser wavelength and thus write off the special problems associated with holography in the visible wavelengths. The delicate process of making world-class holograms involves a great deal of visual alignment and assessment of potentially dangerous optical systems, which simply cannot be done without sight of laser light.

Before risking damage to your eyes, take advice from a professional group such as LaserMet [2] as to the best eye protection in your own individual application.

Above all, please learn these instrumental parts of the holography process thoroughly with low-power (for example, Class 3R) lasers before becoming involved with more powerful systems; unintentional reflections of Class 4 laser beams, even from diffusing surfaces, are, by definition, potentially dangerous to the eyes. Modern d.p.s.s. lasers offer the opportunity to run at low power levels by software control, and it is really useful to take advantage of this feature during table alignment procedures.

7.8.1 Mode of Operation of a Spatial Filter

In the laser beam travelling on the optical table, there are modes of light which originate from various sources of scatter, such as imperfections in optical components within or external to the laser cavity, and including impurities in the lasing medium, on mirror surfaces or in the air, and which result in rays of light which are displaced from the beam axis. These rays, in the case where they are sufficiently parallel with the axis to remain within the laser beam, will interfere with *primary rays* to create unsatisfactory patterns of diffraction when the beam is expanded through a lens and captured on a screen. These defects often have the appearance of groups of concentric rings in the form of a "zone plate" or other miscellaneous features, in what might otherwise be a uniform area of light.

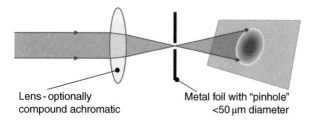

Lens - optionally
compound achromatic

Metal foil with "pinhole"
<50 μm diameter

Figure 7.2 The spatial filter principle. Figure courtesy of Alex Cabral.

The spatial filter comprises a focussing lens, a metallic foil screen with a pinhole of calculated diameter and mechanical means to adjust these components in three dimensions in position with great precision, relative to the axis of the laser beam. The effect of the lens, which is typically an achromatic microscope objective lens, is to focus the beam to a very narrow waist, which, in the fully adjusted condition, coincides with the precise position of the pinhole, as shown in Figure 7.2.

The optics suppliers offer "laser objectives" as an alternative to microscope objectives. These can offer alternative magnification levels beyond the typical microscope objective range of 5X, 10X, 20X and 40X. Whereas these may, in some cases, be cheaper and therefore an interesting option, they are not always achromatic doublets and may be less suitable for *multiple* laser applications.

The laser light which forms part of a simple, single, coherent wave front will pass cleanly through the axially central pinhole, whereas rays which are related to secondary sources are focussed in an annular zone whose light is intercepted by the metal foil.

The emerging cone of light is predominantly free of defects and, upon arrival at a screen, it produces a smooth, featureless circle of light which is ideal as a hologram reference beam, or equally perfect for object illumination. Of course, the absorption of light of any description results in heat accumulating in the thin foil layer, so that "high-energy" pinholes become vital where Class 4 lasers are in use.

Figure 7.3(a) and (b) show the effect of this spatial filter in cleaning up a contaminated laser beam from a helium–neon laser. The photograph (b) differs from (a) only by the inclusion of the foil pinhole; there is a suggestion of shadow in (b) in the position vacated by the dominant ring feature in (a).

7.8.2 Setting Up a Spatial Filter

It is very important to make this task easy by aligning the whole assembly in the first place. The whole mechanical unit *must* be parallel with the laser beam and the beam should enter the precise centre of the objective lens.

In theory, it is possible to argue, since the "launch point" which forms the apex of the diverging conical beam issuing from the spatial filter unit towards an *up-stream* collimator, etc. is the pinhole itself, that it is a consideration to *use the pinhole as the fixed component* and adjust the lens position appropriately to transmit the beam through the fixed pinhole.

However, in practice, for the optical systems used in holography, *the opposite approach is sensible*, since the movement of the pinhole during adjustment is relatively insignificant; no more than a few millimetres in each of the *x, y* and *z* planes.

(a) (b)

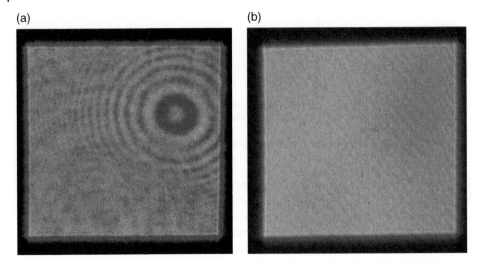

Figure 7.3 (a) Unfiltered spread; (b) after filtration.

(a) (b)

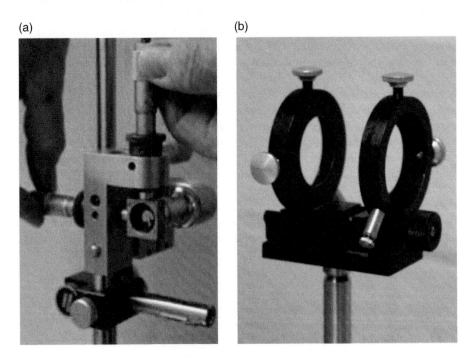

Figure 7.4 (a) Magnetic-type spatial filter; (b) ring-type spatial filter.

Using the pinhole as the mobile element of the filter, we can very easily align the axis of the unit with the laser beam.

The two types of spatial filter which are commonly available are the generic types shown in Figure 7.4(a) and (b).

In Figure 7.4(a), the microscope objective lens is held firmly and the whole unit is aligned with the laser beam, whilst the magnetic mount housing which contains the foil pinhole is adjusted in the x, y and z axes to find the focal point of the lens. In Figure 7.4(b), the rings which hold the lens and pinhole mounting disc appear similar, and offer the opportunity to interchange the two elements. There are extension tubes available which allow you to use lenses of exceptional focal length when the need arises.

7.8.3 Selection of Pinhole Diameter

Any pinhole of 50-micron diameter or less will make an excellent improvement to the beam quality issuing. The reduction of the pinhole diameter towards the minimum possible diameter will offer further improvement of the beam quality, but the minimum diameter is dictated by the focal length of the lens and the beam diameter. A lens of short focal length (high magnification objective) will provide a narrow waist and allow a smaller aperture to function, close to the end of the compound lens.

Many technicians seem to set a difficult target for themselves by ambitious selection of the absolute minimum pinhole diameter. In general, for our purposes, to clean a beam sufficiently to provide an invisible carrier (reference) beam for correctly exposed holograms, the 20X, 10X and 5X magnification microscope objectives can, in practice, be filtered adequately with 25 μ and 50 μ pinholes. The skilled technician can use smaller apertures for special purposes where necessary, such as in a permanently set reference beam for contact copying, where ultimate beam cleanliness is critical.

The practicalities of colour holography mean that spatial filtration of a white beam is far less forgiving; there will inevitably be slight differences in focal length for red, green and blue light, so the advantage of a slightly larger aperture is an important consideration.

"High-power" pinholes mentioned above are made of alloys (molybdenum based) which have high conductivity and high melting points. The focussed beam of a blue or green laser of 200 mW or more is capable of damaging a regular metal foil aperture if the technician allows the focussed beam to dwell on the foil whilst out of alignment.

7.8.4 Aligning the Spatial Filter in the Laser Beam

Whichever mechanical spatial filter type is in use, the first step is to ensure its axial alignment with the laser beam and to ensure its precise height and horizontal position with respect to the beam axis.

The laser beam is reflected, to an extent, from the first surface of the microscope objective lens (which is usually a compound lens). This back reflection can be used to ensure that the lens itself is perpendicular to the laser beam and that it is central, as shown in Figure 7.5.

To do this, the back reflection must be aligned closely with the source beam at a distant optic by holding a card screen close to the incoming beam to catch the reflection a few millimetres from the source beam; you should stop short of allowing the reflected beam to return to the laser aperture where it may well do more harm than good!

The first and second surfaces of some lenses will provide separate back reflections, often of varying size, and the merging of these can be used to sense the precise alignment of the lens. Often, miniature "zone plate" ring features of the type shown in Figure 7.5 will appear spontaneously in either of the transmitted or reflected beams when the correct, on-axis orientation and position of the lens are achieved.

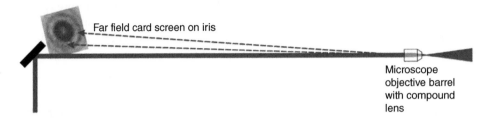

Far field card screen on iris

Microscope
objective barrel
with compound
lens

Figure 7.5 Alignment of a spatial filter lens. Figure courtesy of Alex Cabral.

7.8.5 Centring the Pinhole

Ensure that the pinhole is a high-energy type if a Class 4 laser is in use; reduce the power where possible during alignment. Wind the lead screw system to move the pinhole holder away from the lens to avoid any contact which might damage the lens, and then insert the pinhole mount into the holder, which may be a circular sprung three-point mount or a magnetic block holder, as shown in Figure 7.4(a).

Imagine that the lens will focus the laser beam to a waist no more than a few millimetres from the lens and thereafter the beam diverges, as shown in Figure 7.5, so the displacement of the pinhole to a point which guarantees not only that there is no potentially damaging surface contact, but that the diverging beam provides a circle of illumination which easily covers the tiny aperture of the pinhole, is the ideal starting point for setting up the spatial filtration, as shown in Figures 7.6(a) and (b).

Now, it is important to remember at all times the health and safety aspect of what we are about to do. ***Do not be tempted to look into the pinhole from the exit side at any stage of this process;*** beam visualisation is achieved solely by reflection from a diffuse grey screen viewed from a suitable distance.

Using a screen of grey card close to the pinhole, find the circle of light, as shown in Figure 7.6(b). Provided the pinhole is relatively close to the central position and the objective lens is wound back from the pinhole, as described above, it should be the case that it is relatively easy to find this spot of light.

If this is elusive, close the shutter on the laser beam and reset the pinhole approximately central. Wind away the lens to confirm that the focal point of the objective is not in the plane of the pinhole; if this happens by chance to be the case, then searching for the position of the pinhole that coincides with the ~20 μ laser dot is literally like the proverbial search for a needle in a haystack!

Moving the focus back from the pinhole, however, will allow light to penetrate over a wide range of positions, and allows the technician to wind the adjustment screws of the pinhole position to locate the brightest spot.

Once the maximum transmission is located for a spot, as shown in Figure 7.6(b), the lens is then moved slowly towards the pinhole. The spot becomes brighter, but gradually misaligns with the aperture, with the effect shown in Figure 7.7. Taking guidance from the pattern as to the direction of the misalignment in the x–y directions, move one of the adjustment screws on the pinhole holder to centralise the pinhole in the beam.

That done, move the pinhole closer to the lens again and observe the changing direction of misalignment manifested by the pattern on the screen. "Walk" the pinhole back to the centre of the beam with the corresponding adjustment screw.

Figure 7.6 (a) Positioning the pinhole beyond the focal point to locate the focus; (b) photographic example of locating the focal plane. Figure (a) courtesy of Alex Cabral.

(a)

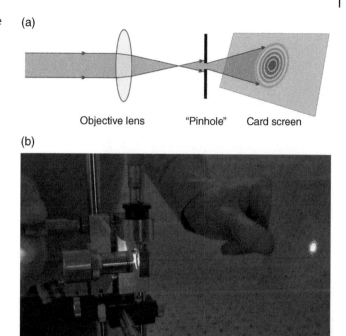

Objective lens   "Pinhole"   Card screen

(b)

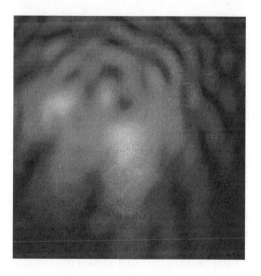

Figure 7.7 Vertical misalignment of pinhole.

The beam on the detection screen is becoming brighter and more perfectly circular. At the point where the pinhole is in the focal plane of the objective lens, we can expect to see a circle of clean, featureless light on the screen, as shown in Figure 7.8; when the lens finally goes too close to the pinhole, so that its focal point effectively falls beyond the screen, the edge effects will again appear, as shown in Figure 7.9.

Figure 7.8 Focussed filter.

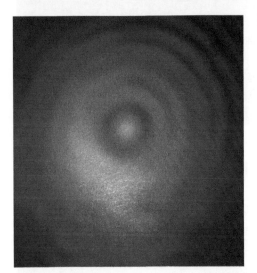

Figure 7.9 Beyond the focal point.

Marks or damage to the lens which produce highly defined defects may not be filtered out. You can tell if these marks are on the lens or in the beam by rotating the lens slightly by loosening the mounting thread by part of one turn.

Discard or clean a damaged lens – dust and fingerprints are often easy to clean off the front surface of the lens with an air duster or lens tissue dampened with solvent. Dry tissue is less likely to soften and remove residue, but is far more likely to scratch surface coatings. Dust trapped within the inner housing is more difficult to repair, but dismantling the barrel and cleaning the other surface of a compound lens is sometimes successful, if the compound lens is dismantled very carefully, taking note of the re-assembly process.

7.9 Filtering a "White" Laser Beam

A topical subject in this era of full-colour holography is the filtration of a "white" laser beam comprising RGB components. When the precautions described later in Chapter 8 for the alignment of multiple laser beams to produce a white beam have been followed

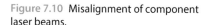
Figure 7.10 Misalignment of component laser beams.

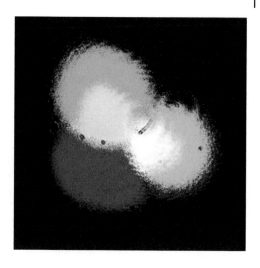

carefully, there is often a disappointment to follow, when the spatial filter is added to the white beam to produce an area of white laser light. Figure 7.10 shows the effect which results from a very minute axial misalignment of the component laser beams. The beams may appear to be coaxial, but magnification by the objective separates these beams into separate circles of unfiltered laser light.

With a little further alignment attention to superimpose the three beam axes precisely, however, the three circles will finally merge into a predominately white circle. At this stage, a pinhole can be added at the focal point of the lens and will successfully transmit all three colour components. Nevertheless, due to differences in the source beam diameter of the three component laser wavelengths, and any residual misalignment of the component beam axes, the spread diameter of each component colour will tend to vary, and its positional limitations mean that the perimeter area will rarely appear as a smooth and orderly colour combination. However, as ever, it is expected that the most advantageous zone of the beam will be the central section, perhaps two-thirds of the overall diameter, and this is the only part of the beam that we should intend to use for object illumination or collimation as a reference beam.

7.10 Collimators

One of the defining components of the optical table is the collimator. Some early manuals of holography practice tended to treat reference beam collimation as a luxury in view of the elevated cost of good-quality mirrors and lenses capable of producing a suitably parallel beam. In fact, in early texts, both Saxby and Benton recommended the "DIY" adventure of manufacturing holographic optical elements to fill this function. Ironically, with the latest developments in the quality of recording materials for holography, the idea of creating "home-made" off-axis collimating optics is now far from impossible.

However, for the production of precision holographic images in a commercial environment, there is no substitute for an accurately collimated reference beam.

There is an option of using mirrors or lenses for the purpose. In practice, there are several considerations to take into account, but the availability of space on the table is often the decisive factor; achromatic behaviour of mirrors is another.

Figure 7.11 The choice between lens and mirror collimators.

The effect of using a mirror collimator is to fold the beam path, whereas the use of a lens will require access to both surfaces and, ideally, a linear configuration. Figure 7.11 shows collimators of the mirror and lens type on a "breadboard" table. These optics are an enhanced aluminium-coated 250 mm diameter mirror and a 125 mm diameter plano-convex lens with AR-coated surfaces; the latter probably represents the largest collimator typically available "off-the-shelf". But this lens is not "achromatic".

As one of the largest components on an optical table, the collimator often has a special significance in the optical layout design, since the diameter of the lens or mirror and the dimensions of its support stand will often dictate the minimum possible height of the optical axis of the layout.

This optical axis height is a really important feature of the optical set-up, because it will dictate the characteristics of the stands and posts which are used to support all other components. The post diameter must be proportionate to the height of an optical component, and the means of attaching the post to the table, damping qualities of the posts and the style of the bosses used to hold the lens holders, etc. are features which ultimately play a decisive role in defining the effective limits of stability of the whole table assembly.

In other words, these are one of the defining factors in the expected exposure times which can be made successfully on the table without undue concern for stability and the well-documented problems of movement or motion marks in the holographic image.

7.10.1 Mirror Collimators

In the event that a mirror collimator is chosen, there are two fundamental types:

1) A spherical mirror.
2) A paraboloidal section mirror (or parabolic mirror).

Consider a collimated beam parallel to the optical axis of the mirrors. In the case of a collimated beam on a spherical mirror (parallel to the axis of the optical elements),

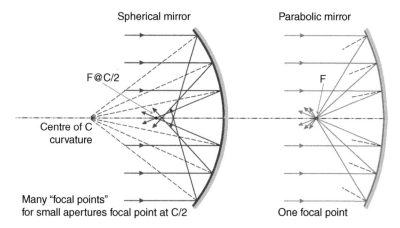

Figure 7.12 Comparison of reflections from spherical and parabolic mirrors. Figure courtesy of Alex Cabral.

parallel light rays that reflect from the central region focus farther away than light rays that bounce off the edges. This is called *spherical aberration*, and its effect may be considerable for smaller focal ratios ("f-number") (< f/6), acceptable for medium focal ratios (f/6 – f/8) and negligible for longer focal ratios.

Take these numbers only as a rule of thumb. In comparison, a parabolic mirror does not suffer from spherical aberration, as depicted in Figure 7.12.

A spherical mirror is ground relatively easily from the pre-form and the cost will reflect this, whereas the parabolic section is, in general, considerably more expensive to produce. Clearly, either type can be coated with the most suitable reflective surface.

In general, we are using relatively low energy levels at the surface during use, but it is really important to remember that during the setting up and alignment processes it is very likely that an unspread laser beam will impinge upon the surface in the centre of the mirror – the very area that we wish to protect from even the slightest damage. I speak from experience when I say that a very expensive gold coating specified as being almost 100% reflective for ruby laser at 694 nm can be irretrievably damaged by the incidence of an unspread argon laser beam at medium power levels. So, yet again, the holographer needs to think of all eventualities *in advance* of experiencing the problem!

In terms of optical set-up, an on-axis configuration is not desirable in holography, as the focal point (i.e. the spatial filter) would be on the axis, shadowing the reflected collimated beam. The solution is to use an off-axis configuration capable of receiving the incident diverging beam from the spatial filter at a specified angle. Then it is, of course, vital that the mirror is mounted on the table in the correct plane of rotation. In general, the axes of incidence and reflection are required to be horizontal on the optical table. This type of mirror is shown schematically in Figure 7.13.

As seen in the diagram, it is basically a smaller part of a symmetrical paraboloid, which makes its manufacture more difficult and therefore implies a higher cost. But its aspherical manufacturing process offers an obvious optical advantage which must be weighed against the additional cost.

It is also possible to use an off-axis configuration in a (partial) spherical mirror but the collimation will not be so perfect, and, to use the rule of thumb mentioned previously,

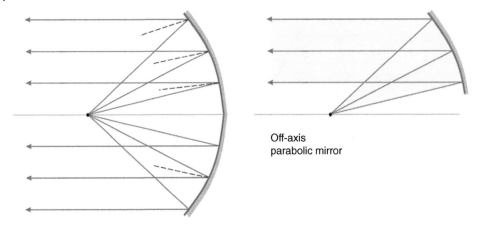

Figure 7.13 The off-axis parabolic mirror and its origin. Figure courtesy of Alex Cabral.

the focal ratio should therefore be determined considering an aperture twice the distance from the optical axis to the limit of the mirror.

It is important in two-generation H1/H2 holography to avoid imperfection of the collimated beam. In fact, such imperfection will cause unexpectedly serious problems in the holographic environment, because the lack of parallelism of rays reaching the master hologram will be exacerbated when we flip the master plate in the plate holder after processing, ready to produce the real image for transfer to the next generation.

This is because of the mis-match of the direction, uniformity and parallelism of rays *producing* the fringe microstructure of the recording at any point in the surface of the master, against the modified direction of rays *reconstructing* the image from the microstructure, at *that same point* in the surface, after the plate has turned through 180°. A common result of such defects is that the final holographic image will sway from side to side when a hologram is tilted or the viewer moves their position.

In practice, the holography studio is able to function well with the cheaper spherical section mirror, provided attention is paid to achieving the best possible alignment. Such a mirror of, say, 30 cm diameter will normally have a focal length in excess of 2 m (therefore a focal ratio of approximately 7). Even in small tables it is possible to use such large focal lengths by folding the beam with a flat mirror, as shown in Figure 7.14(b). The only disadvantage of adding an extra reflective surface is the incorporation of beam defects associated with dust on additional surfaces. But with clean optics and proper precautions against dust precipitation, these defects can be minimised.

With such a large focal ratio, it is possible to rotate the mirror on the table to allow the incident and reflected beams to form a minimal angle of just a few degrees, whilst the spatial filter remains outside the line of the collimated reflected beam as it travels towards the master plate holder, as shown in Figure 7.14(a).

With this configuration, we can achieve minimal levels of image distortion in a hologram, which allow graphics registration in the final H2 to reach levels of precision that are acceptable in both the commercial (security hologram) and display environments.

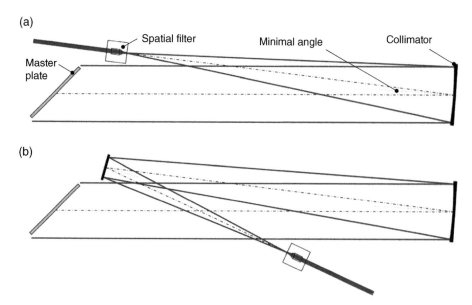

(a) Master plate — Spatial filter — Minimal angle — Collimator

(b)

Figure 7.14 (a) Direct illumination of a spherical collimator; (b) folding the illumination beam for economy of space. Figure courtesy of Alex Cabral.

7.10.2   Lens Collimation

Lenses have the advantage that they may be incorporated in a direct beam path, relatively close to the master hologram *on axis*, with the elimination of the associated problems of coma. Conversely, in the current era of colour holography, lenses are subject to chromatic aberration, which does not afflict mirrors. Lenses of this type are generally of the bi-convex or plano-convex type, but achromatic doublets may well be prohibitively expensive.

For large diameters (> 10 cm) it is hard to find "off-the-shelf" achromatic lenses at an affordable price, making the mirror option the preferable solution, because custom-made lenses to provide achromatic behaviour are generally of impractical cost when ordered in small numbers.

In the case of plano-convex lenses, there is, theoretically, a small advantage in facing the planar surface toward the spatial filter source, since the first flat surface will actively refract the laser beam significantly at the first surface and thus divide the optical activity more evenly (reproducibly) between the two surfaces of the lens, as shown in Figure 7.15(a). Conversely, if rays are emerging from the planar surface of the lens in collimated state, as seen in Figure 7.15(b), Snell's Law tells us that there is no feasible refractive activity at all at the second surface. This effect is known as *spherical aberration* and is always minimised if we maintain the *refraction angle* as small as possible (which is the case when we split the refraction activity at two surfaces).

These are subtle effects, but in the race to make an iconic hologram, every little helps! The level of perfection of the collimation of the reference beam during the H1 and H2 stages is a major factor in the ability to create a really stable and stationary image position in the final hologram when the viewing position or the lighting angle is adjusted.

7.10.3 Establishing the Approximate Focal Length of a Collimator

In the absence of knowledge of a precise specification for a collimating lens or mirror, which can easily happen with older key components in a busy lab with many active technicians, it is a useful first step to make an approximation of focal length by simply establishing the position of the focussed image of a distant paraxial object such as a light bulb or window, etc. on a screen, as shown in Figure 7.16.

7.10.4 Finding the Precise Focal Point of a Collimator

The next step in planning the optical system design is to place the collimator so that its issuing column of parallel rays is incident at the correct angle upon the master hologram plate holder; whilst the plate holder is, itself, in a position upon the table which is

(a)

(b)

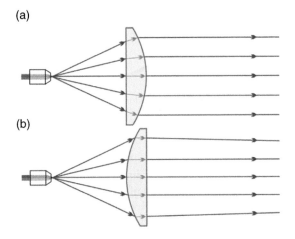

Figure 7.15 Advantage of correct orientation of a plano-convex lens. Figure courtesy of Alex Cabral.

Figure 7.16 Finding the approximate focal length of a lens.

accessible to the technician, and thus allows suitable positioning and access to the art-work display. In the case of a table which will be used for the H1 and H2 recording, also allow space and access to the H2 for its own secondary reference beam, as shown in Chapter 8.

Once satisfied with the position of the plate holders and the collimator, the laser beam, at sufficiently low power, is introduced as the alignment medium. The laser beam is first arranged to be precisely parallel to the table top. A magnetic base with a pole can be marked with a band of white masking tape around the pole at the approximate height of the optical axis, and then a thin (~2 mm) strip of black masking tape around the pole at the exact height of the optical axis plane for the table. This device can be moved around the table regularly to check the height of each set beam as the optical system is gradually built, so that a planar optical axis develops for the whole system.

7.10.5 Plano-convex Lens Alignment

For a lens collimator, we need to ensure that the lens is central, on-axis and is not tilted or skewed in the beam. Putting the lens at approximately the correct height in its stand at the selected position on the table, we then use the unspread laser beam to ensure that the back reflections of the front and rear surfaces of the lens coincide spatially. Even for an anti-reflection (AR) coated lens, the reflections of the laser beam remain easily visible for this purpose.

For a plano-convex lens, the curved and flat surfaces will provide separate reflections which manifest as two dots of different sizes. If you move the lens up and down and side to side, these dots will separate. But when the laser beam falls on the lens at the central position, and the back reflection from the planar surface indicates that the lens is perpendicular to the beam by returning on axis towards the laser source, it is possible to assess the correct height and lateral position of the lens with great accuracy, as the front and back reflections coincide at the central position. Do not allow these back reflections to re-enter the laser aperture; arrange for them to be displaced by a few millimetres from the laser exit port. Light re-entering the laser cavity can cause serious instability, and in the case of a pulse laser, can cause major damage to ruby or Nd:YAG rods, with severe cost implications.

Once the collimating lens is in the correct position with respect to the "raw" laser beam, it is possible to incorporate the spatial filter which will act as beam spreader. In Section 7.10.3, we established the means to locate the approximate focal length of the lens. Set the spatial filter in this approximate position and arrange for it to be approximately paraxial, as described previously. The effect is to spread the beam over the collimator, and an objective lens is chosen to cover most of the collimator with light. We can make use of the reversibility of the light path by placing a plane mirror or flat glass plate in the nominally collimated beam which is emitted from the collimating lens. The light reflected from this mirror or glass surface should be carefully directed back through the collimating lens towards the spatial filter.

Using a small grey screen, it is then possible to determine the plane of focus of this reflected beam. Due to spherical aberration, the screen will tend to capture a cusp of light which appears to rotate as the screen is moved back and forward through the "best" focal point of the lens. At this midpoint, where the image spot is at its smallest, we can now detect the point where the pinhole of the spatial filter should ideally be placed, so the filter assembly is moved slightly backward or forward to achieve this, until we have the focussed spot at the exact place where the pinhole is anticipated to be, after the adjustments are complete. (In practice, we will expect to see this focus on the foil of the pinhole in the final situation, after all adjustments.)

Having established the final exact position required for the spatial filter, it can then be precisely aligned and adjusted, as previously described in Sections 7.8.2–7.8.5, before locking firmly to the table.

It is also possible to acquire a "shear-plate" collimation tester to confirm the parallelism of the rays reaching the master plate. This is a device available from optics suppliers where a fringe structure created between the surfaces of a wedged plate is focussed onto a diffuser screen. The fringe pattern rotates slowly as the divergence changes and the device is marked with an indication of the fringe direction associated with absolutely parallel incident rays.

7.10.6 Spherical Mirror Collimator Alignment

The alignment procedure here is essentially the same as described for the lens. Wearing the necessary eye protection, the mirror is moved until the laser beam at suitably low intensity falls in the centre of the mirror. As described earlier, it is vital that the angles of incidence and reflection are minimal. The reflected beam is arranged to fall in the centre of the master plate. If you have chosen to use the Brewster angle of incidence and are using the appropriate plane of linear polarisation, then that angle can be located precisely by rotating the plate holder about its centre line until the reflected light from the plate, examined on a screen or on the wall of the studio, is at an absolute minimum (extinction of reflection).

The spatial filter is placed at the approximate focal point of the mirror collimator, and, as before, a mirror or flat glass plate is used to reflect the beam back towards the spatial filter. At this position, the plane of focus is detected as the centre point of the rotating cusp, where the spot of light reaches minimal dimensions. Having established the correct position for the spatial filter, it is aligned with the beam, as described in Sections 7.8.2–7.8.5 and then locked down firmly on the table with magnetic or screw fixings. Perfectly collimated reference beams will provide excellent image registration and positional stability in the second generation hologram (H2).

7.11 Organising Suitable Plate Holders for Holography

Plate holders are a particularly vital component of the holography origination system, but suitable components are not generally available "off-the-shelf" from the optics suppliers. The plate holder is, therefore, something which is often necessarily custom-made; not least because the choice of master plate size is a free option to the individual holographer.

The role of the plate holder is to ensure complete positional stability of the recording material during exposure. Ideally, the plate, after processing, can be returned to its original position with great accuracy. However, unlike the stands and holders we use for other optical components, the plate holder will be continually manipulated to allow the fixing and release of glass and film recording materials, so its construction must be at once sturdy and precise in its adjusted positioning.

In practical terms, the method of holding a master plate should prevent any vibration of the glass. The very best way to establish this is to tap the glass with a finger nail after fixing. Any vibration or rattle whatsoever will prevent a perfect recording. Tapping the glass at any point on its surface *must* provide a damped thud.

There are many methods to achieve this. The three-point mount is, in general, the key to the ability to return the plate to its original position. In most cases, it is vital that the plate holder is in the form of a frame which allows access to the plate from both sides. With the exception of photoresist surface-relief transmission holograms (whose glass is coated with red opaque iron oxide), it is usual that we need to address the plate from either side by laser, and the ability to view the object or the holographic image through its substrate plate is invaluable.

Given the concept of an accurately machined open frame with perfectly flat surface and studs for reproducible positioning of a plate, we need then to establish a means for actually holding the plate in a stable position.

7.12   Hot Glue – The Holographer's Disreputable Friend

My use of a flat plate holder system, in the form of a hollow machined frame with thin beads of hot glue on all edges to attach the glass to the flat surface of the frame, brought howls of horror from my former colleague at Applied Holographics, Dr David Greenaway, who was one of the world's most inventive holographers, but whose basic educational and career background was in precision engineering.

However, the rationale for my copious use of hot glue in holding resist and silver halide plates in position followed from the experience of early plate holders in use which attempted to hold the plate literally by sheer mechanical force, and were responsible for actually bending the glass as well as causing cracks and edge damage to occur at the points of contact between glass and metal.

The optical power of a master hologram is arguably akin to a glass lens of *very low f-number*. It is generally accepted in optics and photography that such fast (thick) lenses are difficult to produce with sufficient precision.

The awesome ability of holography to produce this similar optical effect in an active layer attached to a thin glass substrate reminds us of the fragility of this situation, and brings attention to the necessary flatness of the glass. The suppliers of "optical flats" openly relate the flatness of the surface of glass to the thickness of the substrate. For example, a $\lambda/20$ surface of 150 mm square is typically supplied on a 25 mm plate. This is simply because *it is relatively easy to bend a thin glass plate.*

In holography, the typical recording plate is 3 mm glass. This is ordinary float glass of unspecified flatness, and it is immediately clear that the process of holding the plate absolutely still with mechanical fastening undoubtedly has the ability to contort the glass significantly, with deleterious effect upon the detailed image projected from all parts of the surface.

With perfect collimation and good control on fringe shrinkage, any curvature of the glass plate remains the principal contributor to distortion of the image, and there could be no better way to exacerbate this problem than to bend the master plate.

We are in the fortunate position of being able to assess the spatial variability of the thickness and flatness of the glass plate with our laser beam, which automatically offers interferometric facilities. The "wood grain" patterns inadvertently recorded in our holograms are variable from plate to plate, principally for this reason.

Referring to the earlier point about the virtues of hot glue, I found that in the absence of properly designed holders in the early years, the action of a bead of hot glue along the plate edge, attaching glass to metal, had a number of virtues, including:

1) The ability to hold the plate dimensionally stable, but without undue force capable of bending the plate.
2) A solid, temporary fixing where the act of setting could offer a level of contraction or compliance.
3) A certain damping effect in its own right, as a thermoplastic.

Against this, the problem of hair-like threads trailing from the glue and the need to clean residual glue carefully and regularly from the critical surfaces are enough to bring a good engineer to tears! Modern, purpose-built plate holders with well-engineered lugs to hold the plate can offer more elegant solutions. Unlike me, David Greenaway had the experience and foresight to predict this in the mid-1980s.

In the modern era, we find that mechanical solutions with slotted uprights allowing the recording plate to go into position whilst sprung firmly against the rear flanges, or a flat plate frame with three-point mount registration and compliant screw clamps to hold the plate firmly in position without excessive stress, are undoubtedly the best solution in the race for stable, movement-free, precise spatial positioning of the master plate.

7.13   Mirror Surfaces

There are many considerations when it comes to the choice of coatings for first surface mirrors. There are two types of coating:

1) Metallic coatings of various types;
2) Dielectric coatings.

These options must be considered for each *individual application* on the table with reference to factors such as laser power, wavelength, durability, angle of incidence, beam diameter, etc.

It is often the case that reflectors close to the laser will receive significantly higher levels of power *per unit area* than those optics on the same table which are downstream from beam splitters, spreaders and even as a result of general divergence of the laser beam (even divergence of, say, 1.5 mrad is significant as the beam travels the table).

Working with a powerful (> 1 J) pulse ruby laser, for example, there is good justification for using enhanced aluminium or gold metallic coatings *after* the beam has been split and spread. However, at the first mirror (typically at the first corner of the optical path on the table), it is vital to cut all waste limits to the absolute minimum and this is best achieved with a *correctly specified* dielectric (non-metallic) mirror. In the case of high-power pulse laser use, the evaporation of metal coating as a result of overheating may have the unfortunate consequence of creating phase conjugate reflection, with catastrophic results for the laser.

7.13.1   Dielectric Mirrors

Dielectric mirrors comprise a high-quality glass surface coated with non-metallic layers of alternating high and low refractive index and thus operate in the same generic way as the Lippmann volume holograms we have described previously.

Unlike the holographic optical elements that we can make in the Lippmann holo-gram format, the layers are coated by deposition methods which naturally result in layers of precisely controlled thickness laid parallel to the substrate. These layers may consist of inorganic materials such as magnesium fluoride (n = 1.38) and magnesium oxide (n = 1.70).

Ironically, because of the ability of quarter-wave coatings to create interference effects which may be constructive or destructive, similar materials may be used to coat anti-reflective surfaces on components such as windows where we wish to reduce surface reflection close to zero.

The term "correctly specified" means that the wavelength, power and angle of inci-dence for a mirror match exactly the manufacturer's specification. Such mirrors will reflect more than 99% of incident light and the reason that this is so vital is that wasted light will convert to heat and, in the case of a ~15 ns pulse duration, sufficient energy can be released to cause dust to fuse into the mirror and even to cause air to ignite and metal to evaporate, with potentially disastrous consequences – as stated previously, not only for the mirror, but as a result of phase conjugation, capable of causing damage to the laser itself. As we mentioned in Chapter 3, pulse lasers tend to introduce a whole new area of safety issues and it is a very wise and necessary idea to learn your basic holography technique with a low-power, continuous-wave red laser.

The wavelength quoted for the mirror allows for the correct layer construction only when the specified angle is used for incidence and reflection; for example, a mirror may be suited for 694 nm ruby incidence *at 45° incidence*. Dielectric mirrors are generally specified either as normal incidence or 45° incidence. At the correct wavelength, there is leeway to vary this angle by, say, 10° without excessive loss of efficiency; but once these mirrors are mounted in a holder on the optical table, the technician must, of course, remain aware of their specification (by marking the holder).

7.13.2   Metallic Coatings

Metallic coatings of gold, silver and aluminium are relatively highly reflective but the metals tend to be soft and susceptible to mechanical damage. Attempts to clean finger-prints or other marks with soft lens tissue dampened with solvent, or even immersion in solvents, are never 100% effective; *the technician's first intention must be to avoid ALL such contact with the surface of optics*. It is useful to cover optical surfaces when out of use for long periods, but in that case you must take static effects into account. Air dust-ers can be used to remove dust, but these must be used with caution; certain types will emit deleterious propellants when inverted, for example. If you use a lens tissue to remove a dust particle, do not touch the active area of tissue with unprotected hands, and *do not apply any pressure*, allow only the loose edge of the tissue to make contact with the surface, as shown in Figure 7.17, to dislodge dust, and then dispose of it with an air duster.

The optics manufacturers are able to enhance the metal surfaces with harder layers to protect against such physical damage. Protected or enhanced aluminium mirrors are

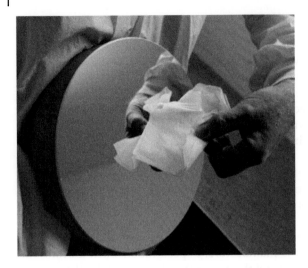

Figure 7.17 Lens tissue cleaning.

ideal for general use on the holographic table. Unlike dielectrics, there is little restriction on angles of incidence and reflection, although acute included angles are always to be preferred.

Mirrors are best held in purpose-made mounts with adjustment screws for fine tuning of table alignment; the "gimbal" type of holder can be adjusted so that the point of incidence does not move spatially when the mirror is moved to align the beam.

7.14 Beam Splitters

These components come in many forms. The primary decision is whether a fixed or variable split is needed for a particular application.

In the case that a splitter is required for low-power work, a choice is available between dielectric and metallised coatings. But high-power laser and pulse laser work may necessitate the use of dielectric coatings, which can have a much higher damage threshold. One obvious advantage of metallic coatings is the tendency to suit any wavelength in the visible range, whereas dielectrics tend to be wavelength-specific (except for certain complex and very expensive multi-layer coatings).

Beam splitters may take the form of a surface-coated glass or alternatively may be a coating upon a right angle prism which is cemented together with a similar blank in order to allow the coated surface to be enclosed within a glass cube. The surfaces of the cube can be AR coated to produce a very efficient product for low-power laser work.

A less well-known type is the pellicle beam splitter, which comprises a thin (~5 μ) membrane stretched over a circular holding frame. These work well for a very broad range of wavelengths, eliminating fringing due to back reflections discussed below in glass splitters, but they do not accept high power and may be susceptible to movement instability due to acoustic or air movement effects. Touching the surface of these splitters may be considered a veritable disaster.

7.14.1 Metallised Beam Splitters

Metallic coatings can be prepared in a graduated fashion. Such coatings are alternatively arranged in a rotary (Figure 7.18) or linear (Figure 7.19) format.

A glass disc rotates about its centre and the graduated coating is organised such that the laser beam passing through the disc close to the periphery passes through an increasingly dense coating of aluminium, with the effect that it is reflected increasingly as the disc is rotated, whilst the transmitted beam is gradually reduced in intensity.

One of the difficulties here is that, as the reflected beam reduces in intensity and the light transmitted through the first surface increases, the back reflection from the glass second surface will tend to create problems, as in Figure 7.19, unless that surface is AR coated. The alternative to AR coating is to use a bigger disc with a thick glass substrate so as to spatially separate the primary beam from the reflections, or, in some cases, to allow the disc to operate at the Brewster angle where polarisation conditions permit. From a thin glass substrate, for example, 2 mm, beware that the reflected beam may also enter a downstream spatial filter objective lens, with serious diffraction consequences for the issuing beam, which, after magnification by the spatial filter assembly, may well contain visible fringes of alternating high and low intensity.

Linear versions of the graduated splitter are also available and tend to be considerably cheaper. Here, the glass plate is held in a linear translation stage.

Both of these types may have a graduation scale attached to the glass, which is useful to return to a previous setting, but these graduations are difficult to set with the required accuracy unless great care is taken, since very small movements of the splitter result in

Figure 7.18 Rotary beam.

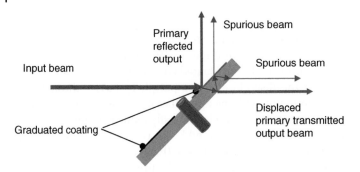

Figure 7.19 Spurious back reflections in a rotary beam splitter.

significant changes of the ratio between the beams given the dynamic of the active part of the gradient.

It is true that there is a lateral gradient of attenuation across a laser beam passing through a graduated splitter, especially if the beam diameter is large, for example where a divergent beam has travelled some distance on the table. One way to improve this is to invert a second identical plate face to face with the original splitter to produce a transmission profile with high density at either end.

Whereas the transmitted beam is directionally stable, the reflected beam tends to wander in lower-quality units of the rotary type depending on the quality of the rotation bearing. This is very frustrating as regards its propensity to move out of alignment with downstream spatial filters or other precision optics.

Remember that the lateral (Snell's Law) stepping of the transmitted beam through the glass plate at 45° orientation will also move the beam axis out of alignment, left or right, with downstream components, as also shown in Figure 7.19, so the beam splitter should be installed in the beam path before those other components go on the table.

7.14.2 Dielectric Beam Splitters

For high-power lasers, the use of dielectric beam splitters offers a safer alternative to metallic coatings, because temperature changes in the metal layer can give rise to phase changes between the transmitted and reflected beams. Naturally, dielectric coatings in their simplest form tend to be monochromatic. However, some manufacturers are now able to produce broad-band coatings to cover a range of, say, 400–700 nm.

The method used to provide a variable beam split is to use a half-wave plate to divide the beam into two polarisation components which are selectively separated by the dielectric coating. P-polarised light is transmitted through the assembly whilst s-polarised light is reflected from the side port. A second wave plate can then be used to return both beams to the same state of linear polarisation moving "downstream" on the optical table.

The dielectric beam splitters, since they are controlled by rotation of an axial half-wave plate, are able to offer precise ratio control as well as convenient avoidance of beam pointing errors, as the splitting component does not, itself, move. Furthermore, for professional installations, they can be operated electronically. However, one requirement of these dielectric splitters is that the dependence upon Brewster separation of the polarised components means that only collimated beams can be separated satisfactorily. The result of an attempt to divide a divergent laser beam will be that the core area of the beam is separated perfectly whereas the Brewster conditions are not fulfilled by the extremities of the beam. The result is that cross-polarised light arrives at the output wave plate.

7.15   Shutters

Conventional electronic shutters are available in which the controller/power supply has a variable delay setting and adjustable opening time. In the modern holography system, it becomes increasingly important to control the shutters by remote computer; often, the control station is outside the laboratory entirely, but in the case of a control station close to the table, delays may be arranged for the technician to leave the room prior to extended exposures. But when setting up a control station outside the lab, remember that you will also need to have control at the table side in order to undertake setting-up procedures and test shots.

Importantly, the shutters used should be "bi-stable". This means that the actual blade used to block the beam is moved, often magnetically, to the "open" and "closed" positions and that, at each of these positions, it is at rest, so that no electric field or residual current is present during exposure. In this way, undesirable heating effects are avoided – because a laser beam passing close to a hot shutter aperture stands every chance of being influenced by index changes in the surrounding air.

Of course, when using high-power lasers, the electrical power may not be the only source of heat. Then, the problem is to dissipate the heat produced when light is absorbed or deflected by a shutter blade itself. It is important to specify the laser power to the manufacturer in order to acquire the correct unit.

Optoacoustic shutters operate quite differently and have the advantage of coping with short, precise exposure times. Their ability to achieve *complete extinction* of the laser beam, however, is questionable and is, in any case, reliant upon absolute precision of alignment. Consider this method where exposures of the order of milliseconds are required.

A similar situation applies to liquid crystal shutters, which are capable of excellent control speed but whose minimum transmittance may be of the order of 0.2% of the incident light.

7.16   Fringe Lockers

In principle, the idea of compensating instability of the table system by electronic adjustment of one of the beams which gives rise to a standing wave is marvellous, and certainly the experience of observing a holographic origination system under the control of such a fringe stabiliser is absolutely remarkable.

The concern is that, ideally, any hologram origination system should have the *fundamental* ability to present a standing wave sufficiently stable for recording purposes without the need for any correction. But when really long exposures are being made on the table, it is reassuring that drift of the standing wave is prohibited in this way.

The mode of operation of the fringe-locking device is to sample a representative interference fringe structure between the relevant beams which are intended to create the standing wave which is the hologram recording. The fringe locker comprises a control system with a piezo-controlled mirror and a detector head which measures the relative intensity of light between two small apertures in the detector surface.

The sampling process comprises the production of a ring of interference between the component beams in a way analogous to the Michelson interferometer described in Chapter 8. A section of this pattern is expanded by a diverging lens to a suitable size such that its period is comparable with the pair of minute apertures in the detector head. Then, if the interference pattern moves, such that the twin detection apertures record a change of relative intensity, the electronic feedback loop is used, in real time, to move a piezo-driven mirror, influencing one of the beam paths in such a way as to correct the beam path length minutely, in order to return the fringe structure on the detection head to its original position.

It is important, of course, that the beam configuration producing the fringe structure is representative of the hologram configuration itself. When the fringe locker is activated, the effect of tapping the optical table in such a way as would normally cause fluctuation of the interferometric fringe structure is eliminated; the pattern upon the detector is seen to be quite stationary despite the attempt to animate it – the effect is breathtaking!

In the case where there is difficulty in arranging for the laser beams producing the hologram to provide a suitable sample at the fringe detector due to limited laser power or difficulties of configuration, it is possible to use a separate laser to travel a similar path on the table, using as many of the primary hologram optics as possible, provided that the piezo-controlled correction mirror is common to both optical paths.

In this way, a very small "tracer laser" such as a 5 mW helium–neon tube, can be used to provide a suitable sampling system for table stability. Thus, it is possible to avoid exacerbating the instability problem by unnecessarily further diluting the primary laser power to create even longer exposure requirements.

7.17   Optics Stands

There is a wide range of equipment available for use as optics stands. Posts of all diameters are available, with various fixings to the table and the optical components, and can be successful provided the optical axis of the table is suitably low on the table. When the axis is high (often dictated by the diameter of a collimator, etc.), then damped posts must be used to steady optics. Connecting the tops of tall posts, particularly with triangular connections, is very effective.

Magnetic bases will allow the technician to set experimental configurations far more rapidly where path length control and axial alignment are required. Once a permanent set-up is devised, however, for example on a permanent mastering system, the posts can be screwed to the table.

7.18   Safety – Reprise

It cannot be stated too often that laser beam manipulation carries a significant danger of eye damage.

It is absolutely vital to discuss these issues with safety experts; take courses in safe handling and wear safety glasses whenever necessary. One of the most dangerous situations as regards eye safety is where more than one person is setting up laser optics. The technician actually making an adjustment is far more aware of what is likely to happen when components are being moved than is an assistant or onlooker; use multiple staff during table adjustment only when it is necessary to do so.

Other safety issues include the following:

- When it is possible to do so, reduce laser power to the minimum necessary level whilst adjustments are made, and return laser power to higher levels only after the completion of alignments. This can be achieved electronically with modern diode-pumped lasers, or mechanically by installing a variable "beam dump" close to the laser aperture. Insertion of a "beam dump" variable attenuator or splitter other than perpendicular to the beam will displace the beam axis in accordance with Snell's Law. It is therefore wise to add this at the set-up stage.
- Use safety glasses to reduce laser light levels wherever and whenever possible.
- Use grey and black cards to trace laser beam paths in place of white cards to reduce diffuse reflection. Any suggestion of smoke from such a surface is an obvious indication that the particular process is untenable!
- When the recommended planar optical path is used, this must be well below eye level; overhead or rising beams present a significant danger to the technician.
- Use plenty of barriers on the table to confine laser beams to limited zones and prevent them leaving the table area.
- Be aware that reflections of laser beams from wrist watches and jewellery are a clear hazard. Tools used on the table such as screwdrivers, Allen keys, etc. present the same risk.
- Install safety interlocks wherever possible to prevent unwanted or unauthorised visitors entering the studio when laser beams are exposed, and generally limit the opportunity for unnecessary visits to the laboratory.
- Fail-safe interlock shutters can ideally be installed close to the laser aperture in conjunction with switching associated with the doors of the light trap. Whereas the idea of switching out gas lasers via the mains supply was unthinkable, the modern d.p.s.s. interlock connection is capable of just that.
- The provision of laser safety warning signs outside laboratory doors is a useful and legally advisable step which often has the automatic and useful effect of discouraging entry by unskilled visitors. An illuminated warning attached to the laser circuitry is an excellent precaution.

Notes

1 Lee Filters, Andover, Hampshire SP10 5AN: http://www.lee filters.com
2 LaserMet http://www.lasermet.com

8

Making Conventional Denisyuk, Transmission and Reflection Holograms in the Studio

8.1 Introduction

Following the ground-breaking work by Denisyuk in the USSR and Upatnieks and Leith in the USA in the 1960s which inaugurated "off-axis" recording simultaneously in the East and the West, typical general configurations for recording the various types of hologram have established themselves, which we will explain in this chapter. The off-axis principle was certainly the launch point for modern holography, as it spontaneously enabled the spurious conjugate images to be separated from the primary image and was therefore arguably the most important turning point for the science of holography, in terms of allowing a whole new ease of viewing the subject matter in both technical and display applications.

Essentially, this was the step which would open up the availability of holographic imaging to applications including display where images would be viewed by the general public. As shown in Figure 8.1, a modern hologram will be configured to receive illumination at an angle of, say, 45° above the normal to the film. This means that the viewer's presence does not conflict with the illumination of the hologram.

In the early work by Gabor and others, where it was necessary to have the "reference" and "object" beams almost coaxial, the shadow of a transparent object (for example, a slide) was featured in the "reference beam" profile, with deleterious effects on image quality, but the methods of Denisyuk and Leith, regarded as being invented simultaneously on separate continents, broke through this barrier.

Nowadays, we have developed some basic conventional generic methods for producing transmission and reflection holograms that each individual user tends to personalise to their own specific needs, and which are detailed below.

In some ways, the technology has developed into two separate parts as a result of the success in the embossing area for security holograms, requiring photoresist masters, and the area of display holography based upon silver halide; the authors have wide experience in both sectors, which we will share in this chapter.

The Hologram: Principles and Techniques, First Edition. Martin J. Richardson and John D. Wiltshire.
© 2018 John Wiley and Sons Ltd. Published 2018 by John Wiley & Sons Ltd.
Companion website: www.wiley.com/go/richardson/holograms

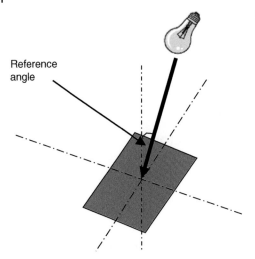

Figure 8.1 Hologram reference.

Reference angle

8.2 The Denisyuk Configuration

The exceptional simplicity of the Denisyuk regime of recording from 1962 is one of its great virtues, but to this day the principle is vitally important and is now the centre of the ongoing race to create full-colour holographic images which literally simulate reality when illuminated appropriately with white light.

A prime reason for the prominence of the Denisyuk configuration is that, in accordance with the Lippmann principle, the hologram is automatically suited to reconstruction with white light; the fringe structure is, in essence, a reflective colour filter, whose operation is in accordance with Bragg's Law, previously discussed in Chapter 2.

From the era of Denisyuk's original work with a single laser, which began shortly after the invention of the first lasers, to the present time, when protagonists of the technique are searching (with considerable success) for "true-colour" image rendition, the fringe microstructure or, in fact, multiple fringe microstructures, are capable of selectively utilising the appropriate colour components of incident white light to reconstruct the original scene. An unfortunate limitation of the method is the spectrum of reflectivity of the object itself, with respect to the laser wavelengths selected for the recording process.

8.3 The Realism of Denisyuk Holograms

The hologram shown in Figure 8.2, made with a ruby pulse laser on the prototype 35 mm film machine, was a key demonstration for the launch of Applied Holographics as a public company into the unlisted securities market (USM) in 1984. The sprocketed 35 mm Agfa Gevaert film was transported through the exposure gate of a small-scale prototype machine. Each incremental movement of the film triggered a pulse of ruby light. But here, the apparent realism is exacerbated by the fact that the hologram is reconstructed at a single wavelength resembling the colour of the original brass badge.

Nowadays, Colour Holographic Ltd is one of the leaders in the field of true-colour reflection holography. The company has recently created holograms of small objects

 Figure 8.2 Denisyuk hologram realism.

including black metal guns which are so realistic that they have been able to challenge visitors to their exhibitions to distinguish between suitably framed versions of the *hologram* side by side with the similarly framed *real* objects.

In the case where these holograms are lit with the original lasers, it is literally impossible even for the experienced holographer (*I remain stunned!*) to distinguish between the actual object and the hologram of it; in some ways this is surely the "Holy Grail" of the science of holography.

8.4   The Limitations of Denisyuk Holograms

The most significant limitation of the Denisyuk set-up, also known as *single-beam holography*, as the latter name infers, is that we are using this single beam to act both as the *reference beam* and as the object illumination, thus providing, by reflection, the *object beam*. We shall see later in other *split-beam* techniques that the ratio of the intensity of the reference beam to the object beam is a critical factor in achieving optimal image brightness.

Now, the main control which we have on this ratio in a single-beam situation in Denisyuk holograms is the nature of the object, with particular respect to the general reflectivity of its surface, its colour, its texture, its dimensional stability, its positional stability and, very importantly, its depolarising effect on the incident plane-polarised laser light used to make the recording. Interference effects are seen only between components of laser beams which share the same plane of polarisation. Thus, light reflected from an object with a change in the plane of incident polarisation will not take part in the process of creating the standing wave required for hologram recording.

8.5 The Denisyuk Set-up

The basic configuration for the Denisyuk method is shown in Figure 8.3. We must consider first the original concept, which relied upon the use of a single laser wavelength with coherence length equivalent to at least twice the distance between the recording layer and the subject matter.

In the Denisyuk recording set-up, rays of coherent light pass through a recording plate, or a film attached to a glass plate, in such a way that a large proportion of the light is incident upon a stationary object, as shown in Figure 8.3.

The object is preferably highly reflective and is best positioned so that its surface tends to direct as much as possible of the light incident upon it back towards the recording material, in order to meet the best beam ratio conditions, as mentioned above. Ideally, the ratio between the reference light incident upon the surface of the recording layer and the light returning from the object will approach parity; for a polished metal surface, this is a realistic target.

As previously discussed, when dealing with the holographic imaging properties of a coherent laser beam, it is convenient to consider the beam in terms of the concept of a wave front, as represented in Figure 8.4. This is the locus of a series of points within a beam which have the same phase; in a collimated reference beam, this is a sequence of planes perpendicular to the axis of the beam; in a diverging reference beam, this is a sequence of planes of spherical section. The reference wave front passes through the recording medium and is incident upon the selected subject matter for the recording.

At the surface of the object, at least some of the incident light will be reflected, as shown in Figure 8.4, its detailed configuration simply a consequence of the physical properties and orientation of the object itself. These properties of the surface and position of the object and the resulting qualities of the reflected light, such as its direction, level of scatter, polarisation and general intensity will ultimately dictate the quality and efficiency of the recording; there is, therefore, every reason to consider very prudently the choice of subject matter for Denisyuk holography – many subjects are at best challenging, and sometimes simply unsuitable, if you wish to produce an engaging or

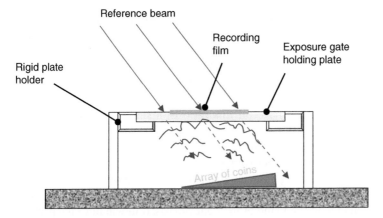

Figure 8.3 Denisyuk configuration.

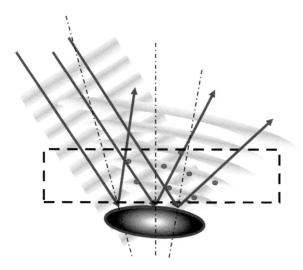

Figure 8.4 Incident and reflected wave front approach.

memorable Denisyuk image. This is because the elusive 1:1 ratio which will produce the brightest image will clearly be reduced both by attenuation of the laser light by the recording layer and the absorption of the issuing beam by the object itself.

8.6 "Recording Efficiency"

For a shiny surface (for example, a coin), the reflected "object wave" may approach description as a plane wave in its own right. Moreover, in the case of a metallic object, the wave front may well comprise predominantly plane polarised light, whose plane of polarisation is common with the reference beam. In this case, one might expect a bright, sharp image: this is probably the reason why the subject has been so frequently used as a demonstration of the potential imaging power of holography for the uninitiated lay-person. But, from Figure 8.4 it is clear that the orientation of such a metallic object is a critical factor in the direction of light reflected, and, in turn, this will strongly influence the viewing properties of the hologram; tilting a coin so that light from the reference beam is reflected normal to the film will always create the most spectacular result. But in the recording of a simple Denisyuk hologram, the ratio between the object and reference beams is unable to fulfil the apparently desirable 1:1 requirement, since the attenuation of the reference beam by the absorbent recording material prevents sufficient light from reaching the object; its own absorption of incident light further reduces the quantity of reflected light which reaches the recording material. Figure 8.5 shows a Denisyuk image recorded with a 532 nm fibre laser. Interestingly, it is easy to recognise the area at the right edge and at the bottom of the image where the coin model escapes the shadow cast by the film itself. The recording film was a small square of film attached to a glass plate supported above the model, as shown in Figure 8.5; the bright areas at the edges are the areas where the marginal reference light illuminating the coins avoids passing through the small square of partially absorbent film before falling upon the coins (i.e. the reference beam left-side ray in Figure 8.3).

Figure 8.5 The shadow of the recording film.

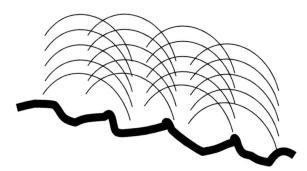

Figure 8.6 Diffuse reflection.

In the case of a diffuse scattering surface (for example, matte white paper), the *object wave front* is effectively a complex aggregate of an infinite number of individual point sources (Figure 8.6). Additionally, in contrast to the metallic subject, the reflected light may be of mixed polarisation or at least only *partially* polarised *with respect to the reference beam*.

Essentially, the complex wave front itself comprises components which interfere amongst themselves. This effect has been termed *intermodulation*. This occurs to some extent in every recording of diffuse subject matter where the subject is relatively close to the recording plate. Each instance of this phenomenon is detrimental to the intended recording of the ideal fringe structure, which ideally would be a *geometrically simple* modulation of the perfectly orderly reference beam with its flat or curved wave front. The holographic recording medium has an immense, but *still limited* capacity for storage of phase information, so that when we record a truly complex microstructure, as one might imagine from an array of multiple secondary sources of this type, it is not surprising that the effective "diffraction efficiency" of the recording is lower than that resulting from a smooth or "shiny" surface such as partially polished metal.

This phenomenon is responsible for many limitations in the efficiency in hologram recording, and there is good reason to believe that it results in the familiar observation that it is relatively simple to produce a "bright" hologram of a *small* object, and very difficult to match that efficiency for a *large*, diffuse object. Like it or not, in our experience, there is rarely a situation where a finished hologram achieves a level of "brightness" which satisfies every viewer.

If the reflected light from the object is, in itself, "shambolic", one has to question the validity of applying the simple term "diffraction efficiency" to the measurement of complex image holograms, since the *recording medium itself* has effectively recorded the *catastrophic* optical situation perfectly accurately!

8.7 Diffraction Efficiency

It is not easy to supply a single, comprehensive definition of the term "diffraction efficiency" (D.E.). When we are considering a simple optical diffraction grating in terms of its ability to deflect a narrow beam of light, we naturally, easily and conveniently revert to the term.

In this case, for a monochromatic light source, D.E. can be considered to be the ratio of light diffracted by a grating into the first order as compared to the power of the incident beam.

As a practical holographer, the subjective assessment of D.E. has often been something of a personal tale of woe. I recall proudly showing the late Dr David Greenaway a film Denisyuk hologram, fresh from the optical table in the mid-1980s, which received the completely unexpected comment, "Oh my God, where has all the efficiency gone?" It looked quite bright to me!

Later, in the 1990s, Applied Holographics MD David Tidmarsh instigated the descriptive phrase "dull as ditch water" when viewing our newly made embossed holograms, in order to encourage us to make ever-brighter, more exciting images. In hindsight, I believe this strategy was successful – holographers are often overwhelmed with pride at the sight of their own new image, and it is therefore wise for managers to assist them to keep their "feet on the ground", because public perception may be unenlightened and sometimes difficult to accommodate, and is often less enthusiastic than the holographer's view. However, "*the customer is always right!*"

As a measure of the "brightness" of image holograms, "diffraction efficiency" is perhaps an inappropriate term, since, in its simplest form, the term tends to relate directly to the ratio of the reconstruction light input to the diffracted "beam" output. But such measurement is not a simple issue, because the output from an image hologram is not a "beam" per se. It is a whole window of illumination containing highlights, shadow areas and intermediate tones, all crafted by the complex grating from a column of incident illumination; and the larger the viewing window, the lower the perceived "brightness".

The casual hologram observer, including the commercially interested corporate Marketing Director (*who has the mandate to order hologram production*) is interested in the "brightness", the eye-catching capability, the ability of the hologram to engage his customer's attention and the *barrier presented to the prospective counterfeiter*.

The diffracted "beam" may take one of a multitude of forms. For example, in general, we could say that a display hologram has a *window of view*, which, in the case of the Denisyuk mode of holography, is equivalent to the recording plate or film itself, but which, in the case of a *second generation* transmission (embossed) or reflection (volume) hologram, is, in effect, the aperture represented by the first generation master hologram. Nowadays, we have an even more complex situation to analyse: not only do we include digital grating arrays in the term "hologram", but most holograms are likely to be multi-channel, multi-colour images, and there is no reason to expect that the efficiency of each colour component or each image channel should be equivalent in "brightness" or direction.

8.8   Spectrum of the Viewing Illumination

For display and security holograms, a really serious new dynamic which has to be accounted for in the age of colour holography is the spectrum of the illumination source. This is a particularly demanding complication in volume reflection holography since the bandwidth of each colour component is a relatively invariable narrow peak of wavelengths. In our experience with full-colour reflection holograms, decisions regarding colour balance, particularly with respect to blue content, are extremely difficult. The tungsten lamp has an evenly distributed range of wavelengths, with its peak in the red area of the spectrum. It has very little ability to produce reflection from a Bragg grating made with a blue laser.

The appearance of a holographic image differs significantly when viewed in sunlight, halogen, fluorescent or LED illumination. Of course, this selectivity of wavelength does not impede seriously upon the viewing of transmission rainbow embossed holograms, simply because the tilting process, which occurs almost automatically in the course of viewing, is able to "tune" the grating to any wavelength peak which is present in the light source. Figures 8.7 and 8.8 show typical spectra for a conventional tungsten lamp and a fluorescent tube, which have radically different colour temperatures. In the case of the fluorescent emission, we see a generally "cooler" light.

Specialist phosphors are made for various purposes, and the effect of these phosphors is to produce variations in the band of wavelengths emitted. Various transition metal ions are incorporated into the basic chemical formulation of the phosphor for this

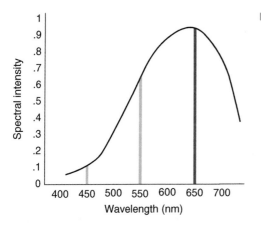

Figure 8.7 Incandescent bulb spectrum.

purpose. The numbered "spikes" in the output shown in Figure 8.8 coincide with the lines we saw previously in the mercury spectrum, but the basic phosphor itself creates a broader band of wavelengths; the shape of the curve is altered by adjustment of the inorganic dope ratios contained within the phosphor.

There is currently a revolution in lighting technology associated with the development of low-voltage LED sources. In the case of the ceric:YAG phosphor type of white emitter, a basic blue LED is used to excite a phosphor in order to produce yellow-green and orange-red light, as shown in Figure 8.9. But there is very little light, for example, at the wavelength of the argon blue line of 488 nm. This means that a reflection hologram produced with such a laser will not appear efficient in this light; but a surface-relief transmission rainbow hologram can be tilted until the brilliant 436 nm line in the source coincides with the linear fringe frequency; unlike the Lippmann hologram, it will appear to be a *bright* hologram.

Figure 8.8 Fluorescent lamp spectrum. Figure courtesy of Alex Cabral.

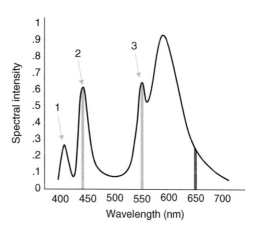

Figure 8.9 "White" LED spectrum.

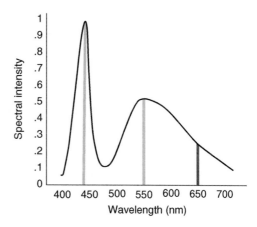

8.9   Other Factors Influencing Apparent Hologram Brightness

As a practising holographer, one is familiar with comments about the inexperienced viewer's perception of the "brightness" of holograms. But this "brightness" is a subjective phenomenon which is entirely dependent not only upon the viewer's expectations, but also upon the illumination of the hologram (including ambient lighting) described above and the type of hologram in question.

Figure 8.10 summarises the relative properties of viewing which categorise the three classical formats for conventional holograms: transmission rainbow (a), reflection H2 (b) and Denisyuk (c).

In part (a), the viewer is effectively looking through a "letter box" at the scene – all of the light diffracted from the source is concentrated at that position, so the eye can be placed at a position of high intensity. Light of various wavelengths is dispersed to slightly different points of elevation. Essentially, in accordance with the efficiency of the grating, some percentage of the total light from the source of illumination is focussed in the narrow image of the slitted master hologram.

In part (b), the viewer sees an image as if looking through a rectangular window in front of the subject. The illumination is reflected in accordance with the efficiency of the grating through a wide range of angles to fill the relatively large window of view. The intensity of light at any specific point of view is therefore likely to be far lower than the concentrated flux in the rainbow hologram.

In part (c), in effect, rays from the light source are diffracted, in accordance with the efficiency of the hologram, through a really wide solid angle so that, in the absence of specular (directional) highlight features, the intensity of light seen from any typical point is relatively low as compared to the methods in parts (a) and (b).

We see from these diagrams an obvious reason why the appearance of the image of a reflection hologram from any single viewpoint appears to defy the fact that Bragg reflection volume holograms are technically capable of higher "diffraction efficiency" than a thin (surface-relief) hologram. The light diffracted from a source of limited power by these reflective volume gratings is dispersed throughout a wide area in two dimensions and so its diluted flux density at any single point is low.

In contrast, embossed holograms tend to appear very bright *only in a narrow band of view equivalent to the position of the slitted master hologram.* Whereas all of the spectral components of the illumination source are utilised by the rainbow hologram, they are diffracted to different vertical positions from a simple, single-slit rainbow hologram.

This factor of "viewability" was one of the driving forces for the invention of the "dot-matrix" holograms, discussed in Chapter 4, whose individual angular plane grating pixels tend to deflect light simultaneously in many directions from various zones of their surface, meaning that, for example, certain parts of the image will be visible to different viewers and will also tend always to "catch the light" when randomly laid on a table in a well-lit room. The individual pixels of these digital holograms are effectively plane mirrors, so their appearance is of extreme brightness over a minute viewing angle, as would be the case for a shard of a broken glass mirror.

So, to summarise, the observation is that any form of hologram has the ability to direct only the light incident upon it, which may be redistributed to a wide range of viewing

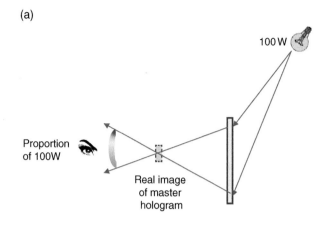

(a)

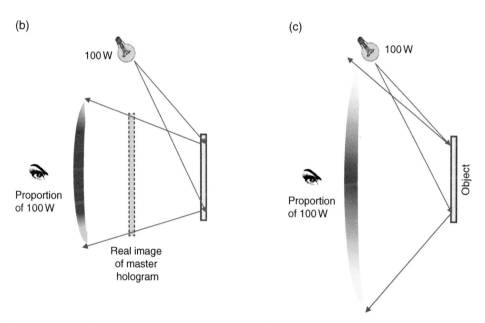

(b)   (c)

Figure 8.10 (a) The transmission rainbow hologram where the "master hologram" is a narrow slit window; (b) the reflection transfer H2 hologram where the real image of the master hologram is generally larger than the final hologram; (c) the Denisyuk configuration where there is little restriction upon the range of positions from which the image can be seen by the viewer, other than the critical angle effect which was described in Chapter 2.

positions, in which case the image will appear relatively dim, or it may be directed to a narrow range of positions, in which case the image, *whilst seen fleetingly*, will appear brighter to the viewer. To reiterate, the perceived "brightness" is inversely proportional to the available viewing angle. Digital holograms (such as dot matrix) attempt to defy this generalisation by allowing varying small portions of the surface to appear brightly illuminated as the viewer moves over a wide range of angular positions.

8.10 Problems Faced in the Production of High-quality Holograms

In all cases where coherent light is used, the minor reflections which always occur at glass/air interfaces are problematic. At the surfaces of the recording plate or film, there will be small levels of reflection, which is undesirable; at the first surface this can be regarded simply as wasted laser light, but at the second surface, it can be more problematic as it will reflect back into the recording material and will, in fact, be recorded as a spurious hologram to the detriment of the intended image. Typically, this type of secondary recording will take the form of a discontinuous surface reflection reminiscent of "wood grain", as shown in Figure 8.11.

This "wood grain" pattern reflects the profile of uneven thickness of the glass; the lack of flatness of *either* surface. This question of flatness of the glass surface can be quite confusing; it is a random variable which can have a significant effect on the aesthetics of the hologram; frequently, there is a wide variation in the quality of the surface, and successive holograms can, therefore, vary in appearance.

In general, in hologram systems, the available laser light must be regarded as precious; every unintentional loss of light will tend to create unwanted reflected beams, which will often create problems, but at the very least, every loss will also have the effect of increasing the necessary exposure time, which, in turn, will exacerbate any problems of instability in the system. For that reason alone, we should always seek, wherever possible, to control and conserve energy at every optical component.

For any simple glass surface in the holography system, there are usually several parameters which we can control in order to optimise its usage and to reduce undesirable back reflection:

- Angle of incidence;
- Plane of polarisation of the laser beam;
- Surface coating for anti-reflection;
- Index-matching couplers.

Figure 8.11 "Newton's rings."

The Brewster principle, described previously in Chapter 2, is much utilised in holography because it allows the elimination of unwanted reflection from a glass surface. To achieve this, the plane of polarisation is organised so that the electric vector is in the plane of incidence (p-polarisation). Then, when the angle of incidence is 56°, for an interface between air and glass there will be zero reflection of the laser beam – all of the light is transmitted.

For Denisyuk holography, however, the need to have the reference beam incident at such a steep angle from the normal (the perpendicular to the glass surface) is very restrictive. One of the problems faced by the holographer in all applications, with the exception of the permanent fixed display, is the difficulty of ensuring that the hologram will be viewed in the intended fashion; that is, with illumination under approximately the same conditions as used in the recording process. It is a practical observation that most casual viewers will conveniently tend to hold a hologram for viewing in the same way that they would hold a printed document (Figure 8.12).

Arguably, a 45° angle of illumination is therefore more realistic than a 56° oblique view. Of course, as Figure 8.13 shows, we are already aware that increasing the angle of incidence results in a greater deflection towards the normal. We showed in Chapter 2 that we do not need to concern ourselves overly with the effect of Snell's Law on the fringe structure itself. But of course, this does not alter the fact that light transmitted through the recording plate or film will return to its original direction of propagation at the second surface before arriving at the object when we are recording the hologram.

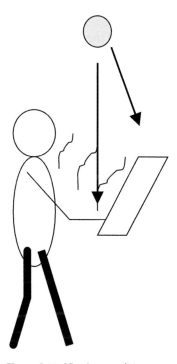

Figure 8.12 Viewing conditions.

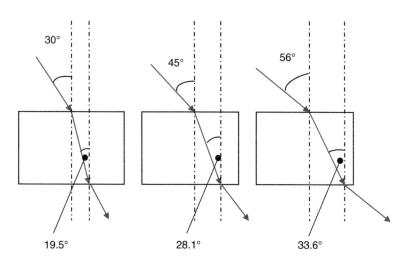

30°   45°   56°

19.5°   28.1°   33.6°

Figure 8.13 The effect of Snell's Law at the recording plate.

8.11 Selecting a Reference Angle

Our selection of "reference angle" is, in the case of single-beam Denisyuk holography, importantly equivalent to the angle of illumination of the subject matter itself.

In the same way as a studio photographer will attempt to use the most advantageous lighting for his subject (for example, key light, flood, diffuser, kicker), it is preferable to provide specially designed lighting for hologram subject matter, but this facility is not realistically open to us in the Denisyuk mode of operation.

If we used the Brewster angle for our reference beam, the subject matter itself would be obliquely lit. This immediately presents obvious problems:

- The reflected light will tend also to be oblique to the surface and may intersect incoming reference light at an unsatisfactory angle. The quantity of light returning from a flat model *in the required direction* towards the recording film may be low unless the chosen object is highly diffuse.
- A surface illuminated at such an angle will tend to have its surface profile interrogated rather harshly. Any returns or indentations will be in shadow; any small protrusions will produce long shadows.
- The surface texture will be critical. For example, from a shiny or smooth surface (polished metal), very little light will arrive perpendicular to the recording layer. A granular (metallic paint or brushed) surface may perform more satisfactorily. Even the *direction* of brush grain may prove critical to hologram brightness.
- The upper edge of the plate holder or the plate itself will cast a long shadow across the object if there is significant depth in the model, so it might be necessary to use a very long plate. Light reaching the object from beyond the edge of the plate will reveal the shadow of the plate, as shown in Figure 8.5.

For these reasons, it is realistic to reject the "Brewster solution" in the design of the recording jig for everyday Denisyuk holograms. If we use a shallower angle of illumination, all of the above problems become a little less critical and the ease of encouraging the viewer to view the final hologram optimally is significantly improved. Light is reflected more directly from a diffuse object and arrives at the film at a more suitable range of angles of incidence. The viewer will eventually view the hologram approximately perpendicular to the recording plate – the hologram has the function of *reconstructing the original object wave front*. For this reason, we must ensure that the original wave front moves predominantly in the direction of the viewer. We should not forget that if that object wave is *not* a visually impressive sight, we are recording exactly that!

So, when the reference beam has been arranged to illuminate the subject matter, before inserting recording material, do ensure that the scene appears in some way spectacular, as you later hope to find in the holographic image. Many non-metallic subjects will generally depolarise the laser light; only reflected rays polarised in the same way as the reference beam will contribute to the hologram recording. Indeed, it is actually worse than that. The cross-polarised light will contribute to the *exposure* of the recording film or plate, but it will not form a diffraction grating; this is rather like increasing the fog level in a photographic recording and will contribute image noise without adding to the image brightness (which is always, as discussed, limited in Denisyuk holograms).

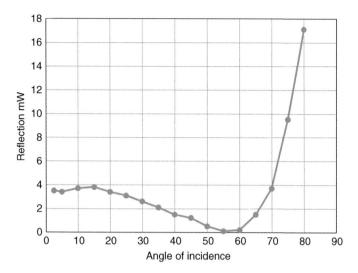

Figure 8.14 Reflection of 50 mW p-polarised laser from glass plate.

Of course, it is therefore always extremely useful to use a linear polariser as an analyser in this situation. If you align the polariser in the reference beam to establish the plane of polarisation, you can then view the subject matter through the polaroid under those conditions of rotation and establish how much of the light reflected from the object is going to contribute to the effective beam ratio.

As discussed above, the Brewster condition cannot easily be utilised to eliminate internal reflections and resulting wood grain effects fully in the Denisyuk hologram.

However, as shown in Figure 8.14, an experimental series shows that there is a clear advantage in the use of p-polarised light even at shallower angles of incidence than the 56° Brewster angle to reduce internal reflections. This is an experiment that is easy to carry out.

In order to reduce spurious reflections in the holding plate shown in Figure 8.3, there are a number of options when deciding how to hold the recording material for a hologram:

1) In the case of photopolymer, the tacky nature of the unexposed film when released from its protective layer will preclude the need for any upper plate. The film will need to be rolled into position with a rubber-coated roller in order to exclude air bubbles. The contact between the holding plate and the recording layer is generally sufficient to eliminate significant internal reflection at the film interface.

2) For glass holograms, use the recording plate itself directly in the rigid plate holder and eliminate any additional supporting glass.

3) Use an anti-reflection coated glass plate or two such sandwich plates to hold silver halide film firmly in position.

4) Add index-matching or surface-coupling liquids to eliminate air interfaces. The capillary effect can often be used to hold film stationary in position without additional measures such as a second glass plate for sandwiching.

5) The idea of a "vacuum chuck" principle is also very attractive for film copying. This is detailed in Section 8.13.

8.12 Index-matching Safety

In the case where liquid couplers are to be used, the two outer surfaces of the glass/film sandwich only may be AR coated with no disadvantage, but longevity is increased as the uncoated surfaces can regularly be cleaned without damage. Then, very small quantities of liquid from a teat pipette may be applied to the surfaces of the recording film. Generally, the recording film can be laid in place with a roller and will remain stationary through surface tension.

Suitable "surface-coupling" liquids for silver halide are alcohols, xylene, white spirit, cedar oil, decalin, etc. As shown previously in the table of refractive indices in Chapter 2, these liquids have a wide range of refractive index values. But since exclusion of air interfaces is the main objective, the exact refractive index of the various liquids, in this case, is not critical.

Subjects like ventilation, skin contact, reagent removal and interaction with the recording material are most important – it is easy to be drawn in to the use of unpleasant chemicals when concentrating solely on index values. Silver halide material will tend to be absorbent to water content of materials, with the effect of emulsion swelling and resulting colour change; photopolymer materials will happily record an image whilst submerged in water! The principle of silver halide layer swelling, simultaneous with exposure, has featured in techniques used in the past by Applied Holographics, Walker and Benton and Smart Holograms.

8.13 Vacuum Chuck Method to Hold Film During Exposure

To avoid the presence of an additional glass plate in a sandwich configuration, it is possible to adopt the idea of a vacuum chuck, which has frequently been used in photographic and lithographic applications in the past. A relatively low level of vacuum is required. A glass, polycarbonate or acrylic holding plate can be used in the exposure gate with a continuous groove cut into its surface, which is evacuated by a tube entering from within the thickness of the plate, as shown in Figure 8.15.

Since the vacuum level can be quite low (certain small pumps of the type used for aquarium tanks have a vacuum facility and are ideal), there is no real problem with movement or vibration, since the central zone of the film and plate is, by definition, quickly evacuated of all air and is thus stationary as soon as intimate contact between film and glass is achieved.

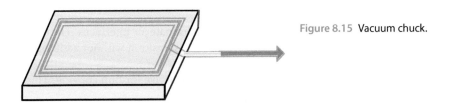

Figure 8.15 Vacuum chuck.

8.14   Setting the Plane of Polarisation

A polaroid analyser can be used to align the laser beam polarisation at the plate holder and should be marked in accordance with the plane of polarisation. Then, it is very useful to view the object *as seen by the reference beam* by looking regularly at the laser-lit object through the polarising filter. Subject matter that makes great holograms in this mode will tend to show a vastly improved image when the analyser is correctly aligned. If rotation through 90° does *not* show a vast reduction in brightness, then the implication is that half the object light arriving at the film will be "wasted"; it will not contribute to the hologram and may well be deleterious to it.

Many metallic subjects such as the ubiquitous piles of coins which are so often successfully recorded in Denisyuk mode, will provide a splendid bright, sparkling, active view through the analyser in laser light. From metallic and other gloss surfaces, almost all incident light is reflected in the correct plane of polarisation, and the "object-to-reference" beam ratio will approach the ideal 1:1 ratio for recording a substantially planar, high-contrast, volume fringe structure in the recording material.

The Denisyuk "camera" can be arranged optionally with either a vertical or horizontal optical axis. This is a matter of choice for the holographer. A minimum of optical components is required for this type of work.

With the horizontal configuration shown in Figure 8.16, the object will effectively lie on its side and will, therefore, need to be fixed upon a rigid support. Some subjects are difficult to support in this fashion and can be better imaged with a vertical system. When the overhead reference method shown in Figure 8.17 or Figure 8.18 is used, there is freedom to use real objects which include liquid content and other flexible objects, since the subject matter can either lie horizontally or stand upright on the table.

However, raising beams and optics high off the table will often lead to instability. There is also an issue of safety with any beam which rises from the table, and extra precautions must be taken to ensure eye safety.

The laser beam is adjusted for its plane of linear polarisation by rotating the laser itself, or by the use of a half-wave retarder plate. It is relatively convenient to rotate

Plan view of optical table

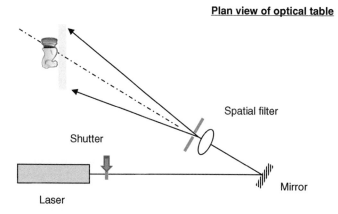

Figure 8.16   Horizontal Denisyuk camera.

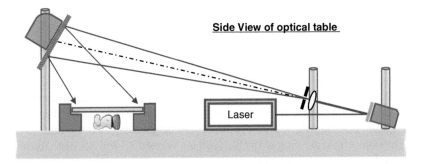

Figure 8.17 Vertical Denisyuk camera with horizontal object.

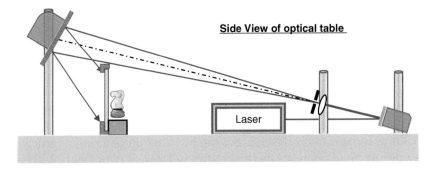

Figure 8.18 Vertical Denisyuk camera with upright object.

lasers such as helium–neon of the tubular type, or modern d.p.s.s. lasers which can quite successfully be suspended on their heat sink in any orientation. Half-wave plates are now available in the form of plastic materials, which have the advantage of being suitable for cutting into multiple pieces. In general, half-wave plates are specific to a given wavelength.

In the case where an overhead reference system is used, the plane of polarisation of the laser is rotated through 90°.

One significant advantage of the process for viewing a Denisyuk hologram is that, unlike the more complex, two-stage ("H1/H2") recordings which we will consider later in this chapter, the single-beam recording method comprises a single, simple recording step which is literally directly geometrically duplicated in the reconstruction process. Our single-laser recording beam, or *reference* beam, interferes with light reflected from the subject to produce a nominally planar grating whose Bragg planes act as a wavelength-selective reflector, and whose general geometric qualities result directly from those of the recording beam. Thus, we can think of the position of the white light point source used for viewing the hologram as being equivalent to the position of the spatial filter in the recording system. Therefore, unless the hologram is intended to be illuminated by a distant source such as the sun, a divergent reference beam is likely to provide a perfectly suitable geometry for a hi-fidelity reconstruction of the original object wave front, which is surely the essence of Denisyuk holography.

One aspect of Denisyuk holography which appeared to have a promising future commercially was the use of ruby laser exposure to eliminate the traditional movement problems, so that three-dimensional models could be recorded for archival storage almost on the basis of conveyor belt delivery.

The US military has previously shown interest in the archival storage facilities for dental casts, which are an important identity aid and take up immense storage space. The holographic records appeared to be a useful substitute for actual casts. The example in Figure 8.19 is a ruby pulse exposure on Agfa film which was chemically adjusted to replay a strong green image for visual purposes. However, if a hologram is recorded at a suitable wavelength to permit viewing with a source of similar colour, then conjugate reference illumination of such images may be sufficiently dimensionally accurate for ID purposes. The casts themselves are made from bite patterns

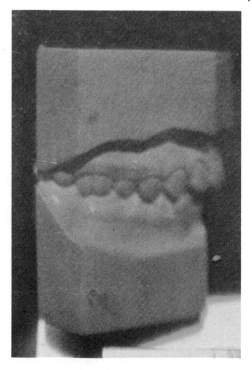

Figure 8.19 Dental cast.

whose pseudoscopic images could be directly reconstructed as an orthoscopic image by a suitable replay beam, thus eliminating the conventional need to produce a "positive" physical plaster cast. Q-switched Nd:YAG pulse exposure at 532 nm, for example, could produce a dimensionally precise image direct from exposure for green viewing ("plaster of Paris" casts are *seriously* diffuse depolarisers).

8.15   Full-colour "Denisyuk" Holograms

As mentioned in the discussion of spatial filtration, the theoretical calculations of pin-hole diameter can set a precise alignment task for the holographer, but in fact increasing the pinhole aperture to twice the diameter recommended in catalogues will do very little to damage the visible beam quality, and will, in fact, permit an achromatic microscope objective lens to focus a three-component laser beam to pass with adequate filtration, resulting in the creation of a "white" laser beam; this is the key to colour Denisyuk holography.

Three-dimensional images of exceptional realism, full colour and wide viewing angle are possible. The image shown in Figure 8.20 shows the potential of this technique. This hologram was made by Jonathan Wiltshire using a Colour Holographic BBV Pan recording plate. It was produced with 100 mW red, green and blue d.p.s.s. lasers. An optical table set-up with filtered laser beams is capable of a rapid turnover of images provided suitable subject matter is selected for the simple imaging process.

Figure 8.20 Full-colour Denisyuk teapot.

The task of filtering three wavelength beam components simultaneously has a number of conditions outlined in Chapter 7:

1) Perfect axial alignment of the component beams.
2) Reasonable similarity of the divergence and diameter of the individual laser beams.
3) Attention to the total power of the component beams. The summation of the power of three lasers impinging upon a foil pinhole can lead to significant heat damage to the aperture. High-power pinholes are available. It is also wise to align the pinhole approximately with the red laser beam only, in the first place, before adding the other beams, so that at no point is all the power of the combined lasers focussed upon the metal foil surface.

8.16   Perfect Alignment of Multiple Laser Beams

In order to combine the output beams of three separate lasers, dichroic dielectric filter mirrors can be acquired, and these are selected on the basis of the chosen laser wavelengths. In this way, the configuration shown in Figure 8.21 may be achieved, where the principal laser beam passes through the uncoated side of a filter and a second beam is

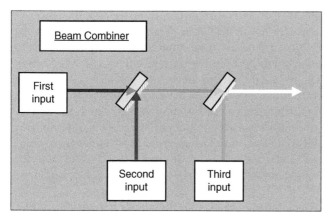

Figure 8.21 Dielectric laser combiner.

introduced at the coated surface such that the first transmitted and second reflected wavelength beams are coaxial *and* spatially aligned.

A second dichroic mirror which is substantially transmissive of these two incident wavelengths is then placed in the beam so as to combine the third laser input to produce a three-component, "white" coaxial laser beam. Absolutely precise adjustment of the angle, tilt and yaw of each combiner optic is advisable; keeping the optical path for a holographic origination system absolutely planar wherever possible is an excellent guideline intention.

It is imperative, not only that the second and third laser beams align in precisely the same direction as the principal (for example, blue) laser beam, but also that their point of incidence is exactly coincident with the axis of the principal beam on the reflective front surface of the dichroic mirror. For example, where the second input beam arrives at the coated surface of the "red mirror", we are interested in the superimposition, *in the plane of the dielectric coated surface*, of the red and blue laser beams.

This is relatively easy to confirm visually, provided the individual laser beams are controlled to suitable levels of power, and provided that safety glasses are used which permit the technician a view of the beams of each wavelength safely and simultaneously.

Use any means at this stage to ensure the best possible alignment – even minor back reflections, including extremely weak reflections of the "wrong" colour from the dichroic mirrors, can help to achieve perfection, and the whole system will permanently benefit from this advantage once set properly.

To ensure that the component beams are, indeed, perfectly coaxial, the white beam is projected into far field and a spatial filter is assembled. Provided that the white beam behaves well at the spatial filter to produce concentric circles of each component wavelength in red, green and blue in the distant location, then even in the event of advanced methods, where we require the beam to be divided by a beam splitter in a position closer to the combiner system, each limb of the split beam can be filtered easily and will provide a concentric display of its component colour circles.

Due to the variation of the Gaussian profile and beam diameter of the individual component laser beams, the "white beam" will tend to display a profile of colour. Telescopic beam expanders can be used to equalise individual beam diameters and to adjust the homogeneity of the circle of light. It is, of course, important to realise that the power of each component laser will influence the hue of the "white" beam and it is vital to remember that the human eye is not the final arbiter of the required balance – that duty falls to the recording material, whose sensitivity to each component colour may vary and is neither necessarily a straight line nor a symmetrical curve.

In order to balance the spectral actinic light level, we have, in the past, installed graduated beam attenuators or beam splitters close to the three lasers, which are able to provide a permanent balance of power for each component laser beam. Additionally, the preferred modern d.p.s.s. lasers can be controlled conveniently from a remote computer. Finally, computer-controlled shutters are used to modulate exposure time for each colour. In the case of photopolymer, the sequence of exposure of each colour component is another consideration.

When shooting a colour Denisyuk in silver halide, our procedure is generally to treat the hologram in the same way as a monochromatic shot to establish first the required laser exposure time and conditions from the point of view of a single colour. Arbitrarily, the red exposure can be used as the first test. You can use a thin strip of glass or film recording material rather than a whole plate to reduce cost. The strip must be sufficiently wide and substantial enough to sit firmly in the plate holder. Tape the leading edge with black masking tape to prevent the reference beam lighting the edge of the strip, because such spurious light will create an undesirable grating as it reflects within the glass sheet.

Mask the strip of recording plate partially and produce a sequence of exposure times by moving the mask across the plate so that a sequence of bands of exposure is produced, in a series representing a suitable range of exposures such as 2, 4, 8, 16 and 32 seconds. Develop the plate as described in Chapter 6 for, say, 60 seconds at 25 °C. Find the exposure time providing a density of 2.0 OD.

It is worthwhile to bleach the plate to ensure that the system is fully functional and at least one of the exposure tests has produced a satisfactorily bright image. Repeat the test for the other colours to determine the preferred exposure for each colour shot.

Then, reducing the exposure time of the first (red) colour to, say, 75%, add the second laser (say green) exposure and balance the image colour by a series of trial exposures by the second laser.

Finally, the final colour component (say blue) is added. Silver halide has the ability to accommodate too high an aggregate exposure by proportionate reduction of the development times; but there are limits to the ability to compensate over- or under-exposure ("process latitude"). Assess results with this in mind after the initial development and bleach processes are complete.

It would be an obvious option to create a procedure where a balanced "white" (as seen by the recording material) standard reference beam was used to record the "true colour" of each Denisyuk subject (scene) in a single combined simultaneous exposure of set time. This would involve balancing the power of the individual lasers by use of software power control or by physical beam attenuators, as previously suggested. But unfortunately, realistically the reflectivity properties of the individual subject matter are likely to require slight variations in the relative exposure power/exposure time/colour balance for optimal recording of each subject.

A series of holograms made with a fixed system would provide a genuine archival record, capable of comparing recorded subjects under the same conditions. However, the situation in "true-colour" holography differs from colour photography considerably.

In the photographic process, a "panchromatic" film records a really broad spectrum of light reflected from the subject matter. So broad is this spectrum that almost all rays of reflected light in a very wide range of visible wavelengths are recorded directly as an amplitude recording in the film. There are slight deviations from a uniform spectrum of sensitivity in the photo emulsion, but essentially, most visible light reflected from a subject is recorded.

In "real-colour" hologram recording, we utilise typically just three lines of fixed wavelength to record the reflectivity of the subject matter with respect to those particular spectral lines. Thus, it is conceivable, and in fact quite likely, that a particular pigment or dye surface coating can have, in its own right, a narrow spectrum of reflectivity which fails to coincide with the interrogating laser wavelength. Hans Bjelkhagen and other workers have proposed that a wider range of laser wavelengths can be used to dissipate this effect, but we rapidly approach the difficulty that the information capacity of the recording material can become exhausted. The recording layer is expected to store independent microstructures of varying frequencies, which amount to complex planar modulations of the refractive index of the gelatin or polymer layer, and these modulation zones are competing for space in the limited confines of the layer. We revert to the problem experienced by Lippmann! We can thus inadvertently enter a paradoxical trade-off between the hologram brightness and its colour fidelity.

As an example, suppose that a cloth backdrop, which appears to be the classical Tyrian purple or Royal purple to the eye, contains a pigment whose low overall reflectivity spectrum presents certain peaks that do not correspond well with the specific wavelengths of our chosen lasers. The balance of reflectivity of the individual components in the "true-colour" hologram and the cloth is likely to be represented in the hologram with a bias toward a red or blue hue, which might not trigger the expected religious or royal associations with which we traditionally link this colour; for this reason, it tends to be more compatible with conventional colour photography.

Phase recording by interference principles has a quite different mechanism to amplitude recording in photography, as we have previously discussed. The use of a light meter on the object wave from a subject illuminated by laser records the "mean" light flux emanating from a complex object and this light will modulate the reference beam to create a standing wave for the recording. This is a quite different condition to the recording of a conventional photograph, where the focussed image from a lens may well include small zones of darkness which simply *do not expose the recording film* and also may contain zones of very high intensity, which, conversely, will tend to cause high levels of exposure in the film.

The challenge to the dynamic range of the recording material presented by holography is often defined by the specular reflections in the Denisyuk mode. For example, in the frequently favoured case of the selection of a group of coins as the subject matter, the specular reflection from the surface of a coin which bisects the angle between the reference beam and the film will reflect the laser beam directly back into the film at a high intensity, which compares starkly with the surrounding zone, as shown in Figure 8.22.

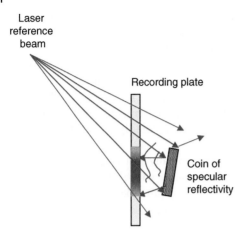

Laser
reference
beam

Recording plate

Coin of
specular
reflectivity

Figure 8.22 The localised "burn out" in a
Denisyuk image.

8.17 "Burn Out"

In Figure 8.22, the zone of recording material shown as a graded brown colouration receives a far higher intensity of total illumination than the remainder of the layer.

Even if the illuminating source was not coherent, the photosensitive film would, to some extent, record a photographic image of some kind in the surface, in the same way that it is relatively easy for a skilled lithographer to record some facsimile of a photograph in high-contrast lithographic film in a contact copy fashion where the film is laid *on top* of the original.

In that case, the steep characteristic curve of the lith film is able to differentiate the areas where the reflected light from white areas of the original print adds to the incident light to exceed the threshold for "infectious" development.

In holography, the presence of such an amplitude-based surface image is often a significant problem. One can imagine also that if the subject of the hologram in Figure 8.22 is, in fact, a very shiny coin, the reflected light will produce a high level of exposure and this may form some kind of amplitude image of the coin on the surface of the film.

When this is slightly displaced from the axis of view of image of the hologram, it can be really problematic; much more so than where, for example, the coin is literally pressed against the surface of the film.

In the latter case, the planar grating which results from the specular reflection back into the recording layer of the reference beam from planar parts of the shiny coin, is effectively a *mirror grating* in the surface whose reflectivity is effectively similar to that of the real surface of the coin, and since, in the contact case only, it coincides with the coin image, it can have a more positive effect.

But if the *specular* reflection from the coin is displaced from the *diffuse* image position of the coin, due to the distance between the model and the film, the interruption of the continuity of the surface by the high level of exposure in a specific small zone is a problem. The presence of a mirror grating and the resulting chemical difficulties in processing such a broad dynamic range of exposure in the film conspire to produce unfortunate effects in the hologram surface, which prevent us from achieving a crystal-clear layer containing a beautiful image that simulates reality.

It is clear that much thought has to go into model design for Denisyuk holograms. Much as the photographer has to consider lighting shadows and preventing undesirable specular highlights, the holographer must solve these problems.

Just like a photographer, we can resort to diffuser spray to "soften" highlights where shiny surfaces are concerned, and one lighting solution for avoidance of problematic shadows leads to a whole area of techniques which may include methods to improve overall image brightness itself.

8.18 Hybrid (Boosted) Denisyuks

Since single-beam holography is limited fundamentally by the inherent beam ratio deficit which prevents the brightness of the image from reaching the optimum, it is possible to use secondary illumination, which will also offer an opportunity to reduce problematic shadowing in the deeper image.

There are two basic approaches to consider. The first is, very simply, the use of mirrors and reflectors at the periphery of the subject matter to reflect "wasted" light obliquely back into the scene, as shown in Figure 8.23. The second is where a beam splitter is used to separate off part of the available laser light to supply secondary lighting to the model, including backlighting of a diffuse rear illumination screen or even a positive photo-transparency background. This approach is shown in Figure 8.24. The beam split from the original reference is spread with a spatial filter to cover the background diffuser, which is optionally covered by a slide (transparency) which contains photographic information. In this configuration, we have a situation where the remarkable realism captured by the Denisyuk hologram is supplemented in an almost surrealistic way by a detailed thematic background, whose emerging light will tend also to act as what photographers term a "kicker" to profile an outline of the subject matter. An artist could consider creating a purpose-made slide to accentuate both effects; the

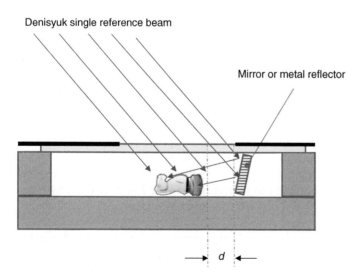

Denisyuk single reference beam

Mirror or metal reflector

d

Figure 8.23 Denisyuk shadow reduction.

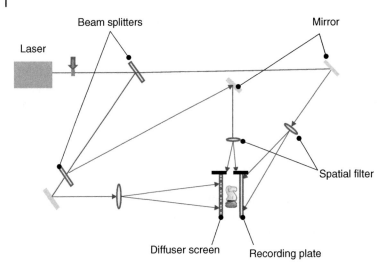

Figure 8.24 Denisyuk boosted by backlit diffuser screen and "top light."

background scene can contain calculated voids to allow pre-calculated illumination or backlighting of the object. The use of an SLM projection on a diffusion screen of this type in place of a transparency will allow "live" visual adjustments of such supplementary lighting.

In order to supplement the brightness and reduce shadowing in any Denisyuk hologram, it is possible to position mirrors or reflective screens at the edges of the frame, as shown in Figure 8.23, to utilise light from the single reference beam which has not been reflected by the object itself; in general, these will be sufficiently oblique to be effectively excluded from view in the final hologram cut to size.

Of course, the return distance from the recording layer via the reflector and the subject must be within the coherence length of the laser. You might decide to have the mirror in view, to include a very interesting reflection of the object, or adjust d in combination with truncating the actual final hologram film to make the mirror or screen itself invisible to the viewer.

The method shown in Figure 8.24 is a greater departure from the pure Denisyuk principle. A beam splitter is used to separate a second coherent illumination beam from the main reference. This is then used to create either backlighting or supplementary oblique object light or both, as shown in the figure. The path lengths must be adjusted to ensure that the light from the laser travels an approximately equal distance from the beam splitter before reaching the recording material on both the reference limb of the system and on the supplementary lighting limbs. This is dealt with in more detail in the following sections relating to conventional split-beam holography. The supplementary side lighting of the object can be achieved by optical fibres which are all lit as one bundle to provide a series of adjustable lights for the object. The path length of the light in the fibres is adjusted (decreased) to compensate the lower speed of light in the dense carrier.

In Figure 8.24, the diffuser screen may be laminated with a photographic image or patterned structure for special effect.

8.19   Contact Copying

Contact copying master holograms or perfectly planar models is also a hybrid of the Denisyuk process. This is a really vital application, since it is the key to the mass replication of film reflection holograms of the "H3'" (hologram 3) generation.

Essentially in this case, the reference beam passes through the recording film with some level of attenuation and is then incident upon a reflection master hologram, as shown in Figure 8.25.

This master hologram is chemically tuned to possess exactly the correct fringe frequency to suit the reference laser beam. Its reflective wavelength is in the form of a narrow distribution whose peak reflectivity coincides precisely with the wavelength of the copying laser, in accordance with the Bragg principle, at the precise angle of incidence of the laser light. This can be a relatively difficult situation to perfect, especially in the case of multi-wavelength ("full-colour") holograms.

Incidentally, although not conventionally under the heading "Denisyuk", the copying of transmission holograms is also possible in a similar fashion. In this case, the film is on the opposite side of the master with respect to the reference beam, as shown in Figure 8.26. In this case, the optimal ratio is not difficult to achieve: the desirable ratio is always less than 1:1 for transmission work, so if the master is less than 50% efficient, this is automatically reached. In this case, the reference beam is partly modulated by the master hologram to create an object wave; the zero order light provides the H3 reference beam, which should be in excess for best results at low noise, thus eliminating the criticality of the need for highly diffraction-efficient masters such as those we seek for reflection contact copying.

One difficulty which may present itself in this mode is a property of certain film base materials which is known as *birefringence* (mentioned in Chapter 2). Birefringence occurs in certain polymer materials such as polyethylene terephthalate ("PET" or "Mylar") which are produced by extrusion or stretching processes, and is due to variations in the alignment of molecular chains. As holographic film is generally coated upon either triacetate or PET carrier, we see the two extreme cases in terms of optical activity, because TAC film with its solvent casting manufacturing

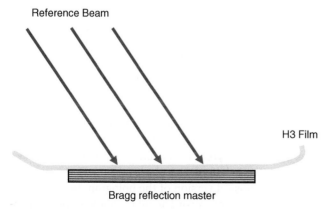

Figure 8.25 Denisyuk principle in the contact copy process.

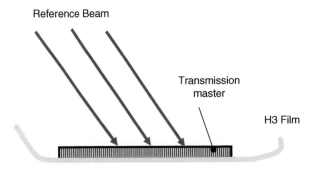

Figure 8.26 Contact copying in transmission mode.

process has exceptionally low levels of birefringence whilst PET suffers quite seriously.

Importantly, the optical activity level is dependent upon the lateral position on the original coated log of film, because that optical activity is created by rotation of the molecular structure, which tends to happen towards the edges of the extrusion, whereas the central section of the coating remains linear in terms of molecular alignment. Therefore, inconsistency from roll to roll results, and may often be traced back to the position a film sample occupied on the coated "jumbo" roll.

8.20 The Rainbow Hologram Invention

One of the great maestros of Holography, Dr Stephen Benton at MIT, gave us an innovation comparable in importance with the "off-axis" inventions of Denisyuk and Leith/Upatnieks when he presented the *transmission rainbow hologram.*

If we wish to make a classical second generation image-planed hologram of a three-dimensional object, whether it is for a reflection or transmission transfer hologram, the conventional first step is to make a transmission H1 master. For transmission hologram purposes, one might tend to use a greater separation between the plate and the object than for a reflection hologram transfer.

This is in order to allow for Benton's intended configuration, where the final viewing stage involves the real image of the master hologram coinciding spatially with the viewer's eyes when the hologram is placed in the intended viewing position. In the reflection mode, the proximity of the master hologram to the model dictates the window of view, and therefore, in general, it is attractive to arrange for the master to be closer to the model to increase the lateral and longitudinal viewing angles.

However, there is a little leeway here and the holographer has considerable freedom of choice. The two holograms in Figure 8.27(a) and (b) show a rainbow transmission and a reflection transfer image of the same model. These holograms were made by Applied Holographics in 1988 from a model by Jim McIntyre. The reflection hologram was made by ruby pulse laser exposure for Graham Saxby's book *Practical Holography* [1].

Later, Applied Holographics used an embossed version of the hologram in a promotional campaign. Whereas both of these holograms record an image of the same model, there are subtle differences in the perspective and the image plane position. The

(a) (b)

Figure 8.27 (a) Rainbow transmission; (b) reflection volume.

reflection hologram was made as an image-planed second generation copy for Graham's book on Ilford film by red laser, but the embossed hologram was recorded by blue laser light as a second generation image-planed rainbow hologram in photoresist. Although less eye-catching in its brightness, the reflection hologram is lower in "noise" and has a much wider viewing window in both the horizontal and vertical planes. In order to achieve this, the first master (H1) is much closer to the model. For the rainbow hologram, the greater spacing of the model from the H1 favours the situation where the ability for the typical viewer to be able to achieve a sequence of single-colour images is more likely, since the slitted viewing window then falls close to the plane of the viewer's eyes when the hologram is held at arms' length in the way expected by the holographer. This is explained more fully later in the discussion of Benton's subsequent development of the rainbow technique.

8.21   A Laser Transmission Master Hologram

The step of producing a first generation H1 master in transmission format is familiar to every holographer, and the general configuration chosen is relatively consistent. In general, optical tables are arranged such that the laser beam paths are planar and parallel to the table top. This means that positional stability of all the optical components is high, since mirrors and lenses can be organised upon relatively short supporting poles. The optical axis is thus a plane which is parallel and as close to the table top as is possible; the minimum height is, in general, dictated by the dimensions of the largest component, which is frequently a collimating lens or mirror whose diameter, as a general rule, should be considerably larger than the collimated beam required to cover the recording plate; only the central part of the beam should be used in precision work. An additional advantage of the set-up described is that it is possible to avoid laser beams travelling the studio at eye level for safety reasons.

The consequence of the planar horizontal optical path is that, in general, there is a requirement to turn the subject of the hologram on its side. A typical set-up for the production of a transmission H1 master recording is shown in Figure 8.28.

The split-beam mechanism means that the holographer now has the ability to control exactly the ratio of the light arriving at the recording plate as part of the reference beam, or the object beam. Within a limited range of values, an increase in the relative intensity of the object light will lead to an increase in the brilliance of the image of that object when the hologram is reconstructed in the conjugate laser reference beam during the production of the second generation (H2) hologram. However, as soon as the optimal level of illumination for the particular object is exceeded, the increase in brightness will be superseded by a rampant increase in "noise" – literally, a halo of light is likely to surround the subject with the effect of a reduction in contrast and clarity of view; such a reduction will inevitably transfer into further H2 and H3 generations of the image. The holographer must adjust the beam ratio until the image is crisp and clear. There is a fine balance between reducing the image brightness to eliminate all noise, and providing a sufficiently bright image to allow a simple and efficient transfer into image-planed H2. The best possible H2 will result from a careful balancing of brightness and noise qualities in both the first and second holograms.

Unlike the reflection hologram gratings mentioned previously, the beam ratio selected for such a transmission recording is much weaker in object light. Test plates should be made for the individual subject to assess the optimum, because it is related to the properties of the subject matter itself. Again, intermodulation is the enemy, but in the case of a transmission hologram, the reference and object light are incident from the same side of the recording plate; the effect of the rays of laser light from the

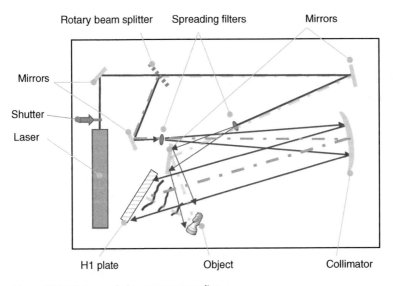

Figure 8.28 H1 transmission master recording.

various parts of the object creating interference amongst themselves is, therefore, even more problematic in producing a bright and noise-free hologram. This is because the frequency of spurious fringes created by interference of rays from distant parts of a large object may well be of the same order as the frequency of the fringes associated with the principal recording and thus highly disruptive of the desired microstructure.

Whereas the ratio between 1:1 and 1:3 is recommended for reflection hologram recording, something of the order of 1:5 (object:reference) is a good starting point for test plates in the H1 transmission mode in silver halide. Typically, we are comparing here a uniformly distributed (specular) plane *reference* wave front with a diffuse and highly Gaussian *object beam* which is typically highlighted and generally unevenly distributed. As the light meter is moved, the object beam varies considerably, as highlights are encountered, but the collimated reference beam is relatively consistent both laterally and longitudinally, with the exception of its predictable Gaussian distribution.

When viewing the processed hologram, variations in perceived image brightness may be witnessed between the centre and the extreme edges of the recording plate as a result of these Gaussian profiles.

Of course, in practice, the holographer must select the most desirable recording condition not only directly upon the basis of the immediate appearance of the H1 master hologram, but, more importantly, upon what will later be the resulting appearance of the second generation rainbow hologram.

With all this in mind, the reference beam must be arranged with a minimal Gaussian variation across the diagonals of the plate; it is a matter of skill and experience to estimate the allowable limits and a level of clairvoyance is advantageous here!

In practice, this requirement means spreading the divergent reference laser beam up to a diameter at the collimator where only the central part of the beam is used, with the effect that the Gaussian profile of the beam is significantly "flattened".

Many early publications, such as that of Saxby mentioned above, and Benton's paper mentioned later in Section 8.25, consider very seriously the challenge of recording holograms without collimators and mention attempts to produce "home-made" holographic collimators, due to the difficulties and cost of acquiring suitable high-quality glass lenses and mirrors for the purpose. However, in the light of experience, I believe that in order to produce first class results, with the exception of Denisyuk holography, it would be making what is a difficult task close to impossible to work without precisely collimated reference beams which provide conjugate illumination for master image reconstruction.

8.22   Laser Coherence Length

In Figure 8.28 we began to deal with "split-beam holography". In order for the laser light from the reference and object beams to be in phase at the recording plate and thus capable of interference between the two, we must ensure that the path lengths of the two beams, from the point at which they are split at the rotary beam splitter to the point at which they merge again at the recording plate, designated by the dotted green and mauve lines in the diagram, are of the same length, or at the least, that they differ by considerably less than the measured *coherence length* of the specific laser in use.

There are two excellent methods to test the coherence length of the laser in use. As mentioned in the section on lasers in Chapter 3, the written specification of a laser with linewidth calculation can become complicated with noise and temperature stability factors, and the ultimate test is a practical assessment of the laser's actual ability to make holograms on the intended optical table.

The classical method for analysing the laser coherence, simultaneously with the positional stability of the optical system and the working environment, is the Michelson interferometer. The original purpose of the Michelson–Morley experiment in 1887 was to explore evidence for the drift of the aether. As previously mentioned, the essential configuration of this historical experiment has recently come to the attention of the public in the gravitational wave exploration. The experiment was intended to show whether these beams would travel at different speeds if the assumption was made that the "aether" was in motion – the recent "gravitational wave" tests [2] reveal an explanation of the Universe which is not so different!

The optical configuration discussed below contains two limbs comprising orthogonal beams of light. We have, in this experiment, an excellent means by which to quantify the coherence length of the light source (again, a problem not dissimilar to the difficulties faced in the LIGO experimentation). In Figure 8.29, the laser beam is incident upon a beam splitter. By carefully aligning the resulting pair of beams to strike mirrors placed at the edges of the optical table, such that they reflect directly back upon the same point of incidence on the beam splitter, the two beams then travel coaxially towards a spreading lens or spatial filter. If these two beam components are coherent, a standing wave of interference is produced in the form of concentric rings which can be examined on a screen. Use micrometer adjustment of one of the end mirrors to centralise its reflected beam in the spreading lens so as to produce the best possible ring pattern on the screen.

These projected rings have several functions:

1) Their existence means that we are seeing interference between the two beams which have reflected from the edges of the table and recombined at the beam splitter before arriving at the spreading lens. If we expand the length of one of the limbs of the system until the rings become difficult to distinguish, then the difference in the lengths of the two beams is a practical measure of the laser coherence length.

2) In the case where the phase of the rings is constantly changing or drifting, we have a measure of the lack of stability of the system. This instability is an appropriate

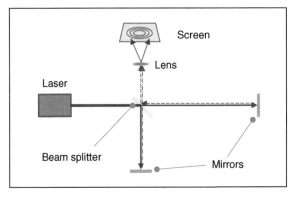

Figure 8.29 Michelson–Morley interferometer.

indicator of the inability of such a system to record a hologram successfully, because these moving fringes represent precisely the type of microstructure of which a hologram recording is made – if the "standing wave" is in motion, the recording material cannot store it faithfully. The ability to visualise this motion provides a good opportunity to explore the whole set-up of a holography studio as well as determining the coherence length of the laser. It is useful, with the help of an assistant, to assess the stability of the optical table and its suspension system as well as the whole environment of the studio and building; try the effects of opening and closing doors in the building close to the table and in other strategic places in the building. I have experienced a situation in the past where the closure of a particular door, quite remote from the optical table, would spoil a hologram recording due to specific resonance, whilst the door of the studio itself was quite inert as regards recording stability!

3) The rings are an ideal manifestation of the instability of the optical system as a whole and can therefore be used to feed a "fringe-lock" system, as previously described in Chapter 7, Section 7.16.

An alternative method to test laser coherence, however, is quite simply to set up a Denisyuk system, as was shown in Figure 8.16. By moving the object away from the recording plate to greater distances, it is easy to find the point where the path of laser light returning from the object to the plate has exceeded the coherence length of the laser. At that point, an image will fade and eventually fail to record, although, of course, the film is still fully exposed. The reference angle can be reduced as necessary for the experiment in order to allow its light to fall upon the object after passing through the plate.

Whereas this method does not provide a live visual demonstration of the stability issues which the Michelson method allows, it does give concrete evidence of the laser's ability to make actual holograms on the particular table and optical system and it can be used with pulse lasers, where visual assessment is not possible. Note that with Q-switch pulse lasers it is necessary to distinguish between coherence length and the actual length of the beam emitted.

If $c = 3 \times 10^8$ metres per second for only 15×10^{-9} seconds, the actual beam of light is only 4.5 metres long, so coherence is capped by that fact.

The "stretched Denisyuk" method above also allows a useful assessment of the power of the particular laser in terms of producing real exposures on various recording materials such as AgHa and photopolymer of the appropriate size for the work intended.

Before investing thousands of pounds in a new laser, it is wise to carry out one or both of these trials.

8.23 The Second Generation H2 Transmission Rainbow (Benton) Hologram

Previous to Benton's "rainbow hologram" invention, second generation transmission holograms were intrinsically unsuitable for illumination with white light. Their fringes, which are predominantly perpendicular to the recording layer, have the property discussed in Section 2.14, whereby they disperse incident white light into its various colour

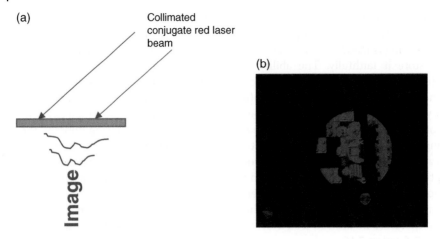

Figure 8.30 (a) Projected real image schematic; (b) photograph of the actual real image projected from a laser transmission H1.

components. The effect of this dispersion is to smear the image of a transmission hologram into a blur of colour in the vertical plane.

There is no better way to examine this phenomenon than to consider a laser transmission master H1 hologram, as described in Section 8.21.

When such a hologram is illuminated by the original collimated red laser beam, we see a perfectly sharp and clear image. If the image is, for example, 125 mm from the recording plate, we can flip the plate through 180° from its original placement after processing to see the sharp, red, pseudoscopic image appear to float in the space in front of the plate, as illustrated schematically in Figure 8.30(a) and as photographed in the form of an actual holographic image in Figure 8.30(b).

However, if we attempt to view this same image with white light illumination, we then experience an almost unrecognisable smear in the image space. This blurred, confused image is difficult to interpret, as shown in Figure 8.31 in the colour photograph of the pseudoscopic H1 real image.

If, however, we illuminate the deep transmission hologram with a collimated *white laser beam* comprising just three single *distinct* component wavelengths, we can see an instructive illustration of the difficulty we face here. With three separate colour components, three images are created, each of which is clearly distinguishable to the eye, as shown in the photograph of the projected image in Figure 8.32. Here, we can see the three image components and we can also see the discrete yellow (red + green) and cyan (green + blue) zones caused by colour mixing.

From such a problematic starting point, Benton's rainbow invention revolutionised holography arguably as much as the earlier off-axis inventions mentioned above, because it specifically gave rise to the embossing technique which has commercialised world holography into a billion-dollar industry.

Benton observed that the colour-smeared white image, which a white spotlight produced in an image-planed second generation transmission hologram, contained a range of views of the subject, which comprised the mixed vertical gamut of parallax recorded in the master (H1) hologram, but that images in each of the various colour components

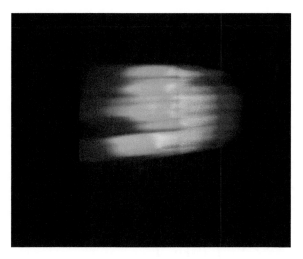

Figure 8.31 Colour-smeared image from collimated white light illumination of H1.

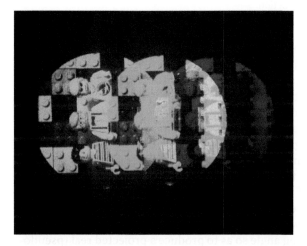

Figure 8.32 H1 image produced by three-component "white" laser" beam.

of the light (a continuum) were displaced vertically with respect to one another in accordance with their wavelength.

He realised that if we could see the individual single-wavelength reconstructions of the type shown in Figure 8.32 *one at a time* in the second generation (H2) hologram, we would have a dynamic new form of hologram, and thus devised a brilliant solution to the problem!

The simplistic representation in Figure 8.33 shows each point on the surface of the dispersive image-planed H2 transmission hologram producing a range of reconstructions of the H1 plate. In a Lippmann reflection hologram, which is effectively a reflective colour filter, only a relatively narrow band of wavelengths reconstructs the master hologram, which is therefore reproduced as a sharp image. The same sharpness is seen when a transmission master is illuminated with laser light from the original laser reference

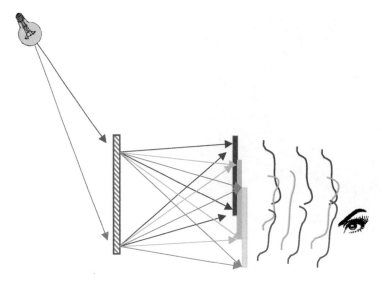

Figure 8.33 Dispersed white-light reconstruction.

beam, as described above, but the introduction of white light illumination immediately causes the dispersion problems which make for unsatisfactory viewing.

Figure 8.33 shows, however, that the image of the H1 master, when a point source of white light is used to illuminate the H2, becomes a continuum of component real images, each of which contains its own secondary image of the original subject matter. Thus, the viewer witnesses a vertically blurred compound image of the original subject matter, of similar appearance to Figure 8.31.

Whereas the eloquent mathematical and geometrical explanations of Benton's concept clarify the mechanism of the technique, I have noticed that many people do not have a clear mental picture of the configuration of the rainbow hologram transfer. The layout is shown in Figure 8.34. The previously recorded H1 master from Figure 8.28 is masked to a narrow slit with black tape and is illuminated from the reverse side at the original recording angle so as to produce a projected real (pseudoscopic) image of the object.

In the past, when laser power was often a major issue in the ability to make transfer holograms reliably, many holographers would use cylindrical lenses to light the H1 master in the second transfer. In this case, it is vital to consider the implications of producing a strip of laser light with the correct geometric qualities. To use a single cylindrical lens to spread a laser beam results in a beam diverging in one plane. Provided the lens has a suitably long focal length, this may be sufficient for simple, small holograms. However, the slit *width* will then be fixed at the diameter of the laser beam.

The other methods to achieve this objective are to use two transverse cylindrical lenses of differing focal lengths, and these can ideally be arranged close to the focal point of a collimator so that the issuing beam is approximately collimated in both axes. Again, the required level of precision of the result is to be considered before investing in the correct quality of the optics; for an object hologram, the distortion associated with imprecise illumination may be quite acceptable, but for colour holography with

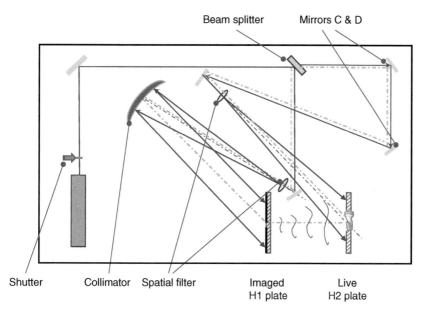

Beam splitter   Mirrors C & D

Shutter   Collimator   Spatial filter   Imaged   Live
                                           H1 plate   H2 plate

Figure 8.34 Rainbow transfer.

requirements for registration of graphics, for security work for example, the utmost attention to beam quality is necessary.

When setting up the H2 transfer of a completed H1, the slitted recording of the object (without vertical parallax) can be captured on a viewing screen to determine its exact position in space. The second recording plate is then placed in the plate holder and its "z" position is adjusted so that the image straddles the plate at precisely the point where the final image plane is required to intersect the surface of the final hologram.

Where a laser of limited coherence is used, the beam paths marked in mauve and cyan must be equal and the mirrors C and D can be moved to achieve this. The space between the H1 and H2 is a part of the mauve beam path that is sometimes forgotten! After so many years of reliance upon beam path correction, in the era of limited coherence length, the authors tend happily to follow the traditional procedure even when lasers of long coherence are in operation nowadays.

Now, in a second generation (H2) hologram, which is, after all, a very complex recording, the "image" of the subject, in the viewer's perception, is effectively contained within the real image of the master H1, and is effectively situated straddling the plane of the final hologram layer. From the viewer's perspective, the real image of the slitted H1 plate itself, "floating" in the space before the final hologram layer, effectively contains a distant image of the object apparently projected back into the film plane of the final hologram.

Benton recognised that the dispersion of colour into its components by a linear surface-relief grating would result in a smeared image of the H1 master, simply because the image created by incident light of each wavelength would appear slightly displaced vertically. In turn, the viewer would therefore see a white smeared (blurred) image of the subject, as we saw in Figure 8.31.

Lateral thinking dictated that reducing the master literally to a thin, horizontal slit would allow the viewer the sight of a narrow slit ("letter-box") view of the subject *with no vertical parallax information*. Thus, when the same "colour smear" effect occurs, it provides a vertical rainbow continuum of views of precisely *the same image, in the same position*, in a range of colours. As the colour changes, there is consistent (zero) vertical parallax and thus no blurring, and vertical parallax information about the subject has effectively been conveniently exchanged for an exciting range of iridescent colours.

This does not disturb human perception, simply because we have binocular vision which is acutely sensitive to parallax information in the plane of our eyes; more specifically, our brain is able to decode the difference between the images perceived by our horizontally displaced eyes in terms of depth information. But the concept of vertical parallax is a minor factor in our perception.

The resulting image qualities of Benton's "rainbow hologram" coincidentally matched exactly the requirements of the security industry – labels which were cheap and provided a very bright, 3D image whose iridescent colours varied with viewing conditions – the optical variable device (OVD) which confounded duplication attempts by conventional copying methods. The rest is now history.

8.24   Developments of the Rainbow Hologram Technique

As holography began to achieve growing popularity as a result of his rainbow hologram, Benton returned to the drawing board to publish his "α-angle" paper – "The Mathematical Optics of White Light Transmission Holograms" [3].

The natural development of the rainbow method was to produce holograms which contained a plurality of fringe structures so as to produce more than one spectrum of colours, with spatial separation of the artwork associated with each rainbow spectrum.

This leads to some confusing terminology, because there is an immediate tendency to refer to "full-colour" rainbow holograms; but obviously, nothing could be more fully coloured than a rainbow! The intention, however, is to create a holographic image which, *from a single central viewing position in the vertical plane*, demonstrates subject matter which is divided into realistic (or otherwise attractive) iridescent colours.

At Applied Holographics in the late 1980s, we had begun to produce multi-slit rainbow holograms and we were also using techniques introduced by Nigel Abraham to produce the commercially successful "Laser-foil" images which involved an ingenious photoresist surface-masking technique to enable arrays of colour-separated embossed colour holograms to be produced with admirable brightness and consistency. The position of the subject matter itself was locked into the surface by the physical position of the masking.

This technique was an incredibly labour-intensive process, which, at one time, involved work where our four holographers, Andrew Rowe, Claire Lambert, Jane Gaffney and myself, worked a shift system to keep the optical table productive for the maximum possible number of hours. The technique produced some remarkably clear and commercially attractive images, such as the novelty label shown in Figure 8.35 [4]. The intense, diffuse colours produced by the method were eye-catching and proved extremely popular until the process was eventually superseded by the dot-matrix technique in the Applied Holographics portfolio.

Figure 8.35 "Laser-foil" method.

However, when the full-frame 8 x 10" resist images from this type of work were step-and-repeated for the embossing process, the problem perceived by the high-security technologists was that the perceived rainbow colour would change at the edge of each image frame from a red to a blue bias.

This was one reason why, for high-security work, the company was looking towards fully three-dimensional colour images stepped on an individual basis, and Benton's brilliant paper, which began by summarising the transmission rainbow hologram as a combination of a prism and a cylindrical lens, certainly inspired the work that Andrew Rowe and I carried out for stereographic and 2D/3D images, including the Glaxo "Zantac" hologram mentioned elsewhere.

Holograms of this type utilising the rainbow principle with the advantage of multiple slits providing colour effects became suddenly popular and, thanks to Benton's α-angle publication, it became feasible to improve the precision of the recording conditions in order to provide the viewer with an improved, simultaneous spatially registered view of the subject in all of its colours.

The description of a transmission hologram as a combination of a prism and a lens clarified the ability to use ordinary trigonometrical calculation to plan the configuration of the optical components and the subject matter in order to achieve the optimal visual experience for the viewer. Contemporary papers by Steve McGrew [5] and Suzanne St Cyr [6] also show how to organise the optical set-up for a rainbow hologram on the basis of work sheets which calculate the "hinge point" – these authors use a geometrical approach rather than Benton's trigonometry, but the result is similar in the practical production of transmission holograms.

For me, the key to Benton's 1982 paper for the Proceedings of the International Symposium on Display Holography, at that time, was the comparison of the rainbow hologram with the combined effects of conventional optics; in the form of a prism to refract and disperse light from the source of illumination and a lens to focus that dispersed light in the line of the viewer's eyes. This allowed the mediocre mathematician to calculate, in familiar terms, the parameters controlling the ability to organise the viewing properties of a rainbow hologram. It also provided a fundamental background understanding of the nature of the rather complex optical device we were creating, and thus readied us to utilise the technique to full effect to provide the full-colour images the security industry demanded at that time.

In his script, Benton described how the conventional prism and lens approach shown in Figure 8.36, with which we could all clearly empathise, described the two principal functions of the rainbow hologram and then went on to say that it was, in fact, more appropriately compared with a linear diffraction grating and a zone plate combination; thus justifying simple mathematical treatment as a pair of diffractive elements, as shown in Figure 8.37.

It is worthy of note that the dispersive effect of a conventional prism causes the blue light ray to rotate more dramatically than red; and that, similarly, a simple convex lens causes the effect known as *chromatic aberration*, where red light is focussed farther from the lens than blue light. Benton's own mnemonic tells us, on the other hand, that "red rotates radically" (the three Rs) when we are dealing with diffractive optics, since the longer wavelength interacts with the grating in a more pronounced fashion, whilst in conventional optics, high-frequency blue light travels more slowly in a dense medium such as a glass prism or lens (see Chapter 2).

In Figure 8.37, we see that the red light from the illuminating source is focussed closer to the hologram plate than blue light and also displaced upwards as compared to the blue light, in accordance with our observation in Figure 8.36.

Benton's idea was simply to acknowledge this fact, and to create a first generation H1 master in such a way that it would lie in this position, as opposed to the conventional parallel plate configuration in Figure 8.34. In this way, the colour image component

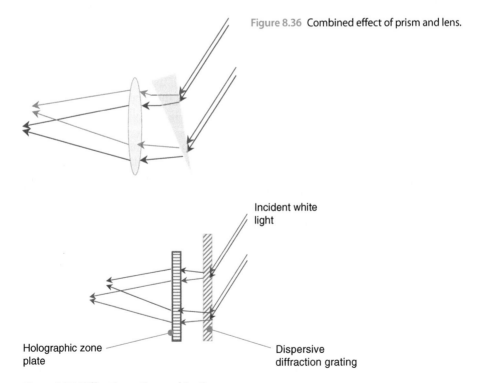

Figure 8.36 Combined effect of prism and lens.

Incident white
light

Holographic zone
plate

Dispersive
diffraction grating

Figure 8.37 Diffractive optics combination.

images in red, green and blue could be recorded in a way that corresponded closely to the exact conditions under which they would later be viewed.

The most important single point is that we are all familiar with the fact that our eye placed at the focal point of a positive convex lens will see the lens filled with light from a distant source (*every reason never to look at the sun through a lens and also every reason to understand why embossed holograms can appear so bright when viewed correctly*).

The rainbow hologram (in a one-dimensional sense) achieves exactly this effect – within the limits of *diffraction efficiency*, the viewer sees all of the light of a certain wavelength which arrives at the hologram surface from the illuminating source, focussed into a narrow linear zone which can be interrogated by both of their eyes.

Thus, we have horizontal parallax and three-dimensional appearance in a single colour of very high brightness. If the viewer moves upwards, the colour changes towards the red end of the spectrum (bathochromic shift) and if the viewer moves downwards, the colour changes towards blue/violet (hypsochromic shift). The slitted master means that there is no vertical image parallax information, so the transition between the colours involves no blurring of the image – a simple, continuous colour change of the single graphic image occurs.

Referring back to Figure 8.28, in conventional style, we can relatively easily record a standard H1 transmission recording. This was the conventional original configuration for recording a first generation master hologram on glass. Because only a narrow strip of the master is required for the transfer of a simple, single-component image to a second generation master, it is possible to use a narrow strip of recording material in rainbow holography.

In certain embossed security work there is no requirement for realistic colour. The "OVD" requirements, in some ways, lend themselves better to the production of multiple random colouration and deliberate switching from colour to colour. Here, there is no limit to the number of individual colours (spectral positions) which can be recorded on a master hologram.

So, the idea to use a whole master plate, as in Figure 8.28, and to divide the surface into various zones of colour (longitudinal) and animation (lateral with respect to the image) is very successful where planar subject matter is used. Only in the case of significant image depth does the problem of mis-registration of colour components begin to show itself, because each image component has been replayed in a slightly different position with regard to the perspective of recording conditions, due to the viewing anomaly demonstrated by Figure 8.37.

8.25 Using the α-Angle Theory to Produce Better Colour Rainbow Images

As a result of Benton's paper, it became relatively easy to predict the optical configuration required to produce aesthetically pleasing multi-colour images and avoid some of the problems of mis-registered colour.

How good was Steve Benton? During his visit to Applied Holographics in the 1980s, I showed Steve my "pride and joy" experimental rainbow hologram of an image projected by an LCD screen. I had borrowed, from Andrew Chapman at ImageDisplay Ltd, an early version of a smectic LCD whose pixels interchanged between diffuse and translucent. I had experimented with both the projected light from clear pixels onto a screen and hologram recording directly of the diffuse pixels of the screen itself as the hologram subject. The rainbow hologram was reasonably good, and I expected a gasp of amazement, but Steve said, "Yeah – we've done a lot of this type of work." Before I could express my fears of laser damage to the delicate LCD screen, Steve said, "and you don't have to worry about optical damage to these LCD screens, remember they're designed to receive hundreds of watts of projector light, not just a few milliwatts." Clairvoyance too!

In our discussion about the possibility that such automation and computerisation of the process might allow us to make more complex images more conveniently, I realised that Benton had made, and would continue to make, an outstanding contribution to many areas of holography, and looked more closely at the work of this brilliant individual.

When I came across the α-angle paper configuration, I immediately understood that the logical conclusion of Figure 8.37 was that the diffractive effect of light reconstructing the final hologram is such that the images formed by the red, green and blue light will be seen at specific, calculable positions whose locus is a plane tilted sharply towards the top of the hologram. This angle of tilt Benton called the *α-angle*. As shown in Figure 8.38, it is related directly to the angle of incidence of the illuminating light in the form:

$$\alpha = \arctan \theta$$

where θ is the reference angle of the rainbow hologram.

Thus, for a 45° (reference) hologram, the α-angle is 35°.

Benton's first logical step, taking the approach of working backwards from the final product, which is a wonderful discipline in holography generally, and assuming that transmission holograms are, in effect, thin holograms, was to use the grating equation to calculate the grating spacing, d, which gives us the central green zone of the rainbow in the horizontal (perpendicular) viewing position, as shown in Figure 8.38.

Thus:

$$\lambda = d\left(\sin\theta + \sin 0\right)$$
$$550 = d\sin 45°$$
$$d = 550 / \sin 45°$$
$$d = 778\,\text{nm}$$

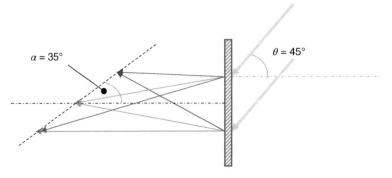

Figure 8.38 The α-angle phenomenon.

The issue of dealing with thick or thin holograms becomes significant in the manufacture of embossed holograms; it is frequently the case that we have two materials on the same transfer table with quite different properties, which are silver halide (a thick hologram with an emulsion depth of ~7 μ) and Shipley photoresist (normally coated at ~1.5 μ and incapable of any significant volume effects). There is absolutely no dispute that the final embossed hologram is, by definition, a thin surface-relief hologram; and this is where we begin to consider the rainbow hologram viewing requirements!

Benton's paper dealt with a hologram made so that its central viewing position (height) coincided with the green view and, clearly, when using a green laser this viewing condition will be matched by the recording geometry where the image is also perpendicular to the plate or film.

But photoresist is sensitive only to blue light and, as the "template" for the metal embossing shim will dictate the fringe spacing, we need to compensate the fringe spacing by adjusting the angle between the beams, which is clearly entirely legitimate for a thin surface-relief grating.

Assuming the use of a 458 nm argon laser, the H2 surface-relief hologram must be shot with an angle between the beams which satisfies the fringe spacing requirements, which will then lead the central (green) hue to be visible normal to the final hologram layer. From the above calculation, we know that we require fringe spacing of 778 nm to diffract green light into a horizontal plane from 45° elevation of our white source.

$$\lambda = d \sin \theta$$
$$458 = 778 \sin \theta$$
$$\sin \theta = 458 / 778$$
$$\theta = 36°$$

For "full-colour" rainbow holography, the essence is to produce three-component gratings which will each provide a spectrum of light displaced in such a way as to allow the viewer to witness the subject in a range of colours. At the central viewing position, we might well attempt certain subjects to make these colours representative of reality – for example, a portrait with lifelike flesh colour, blue eyes and red lips.

Figure 8.39 shows an embossed stereogram filmed by my group at Applied Holographics plc. and transferred to resist by Craig Newswanger at Applied Holographics

Figure 8.39 Tilting a "real-colour" embossed hologram from its central viewing position.

Inc.[4]. The colours at the central viewing position are very realistic, as determined by flesh tones, which are sensitive to variation of colour components and are accurately represented in the hologram at the optimal viewing position. In fact, Figure 8.39 summarises the strength and weakness of the rainbow hologram: if we lower the camera by just a few degrees, the lips of the subjects and the boy's shirt rapidly change to green!

So, with the correctly planned viewing conditions, we have a "real-colour rainbow hologram", which terminology admittedly, at first, appears to provide a curious oxymoron!

8.26   Aligning the Master Hologram with the α-Angle

The "parallel plates" recording method for a transmission master, shown in Figure 8.34, works well for single-slit rainbow holograms and for colour holograms where there is insignificant image depth, non-critical image component spatial registration and no real desire to view predictable key colour components simultaneously. Stephen Benton's rationale for the α-angle work was that the real image of the master hologram would be constructed in a specified, predictable position by the white light used to view the second generation image-planed hologram.

In terms of dimensional and colour distortion, there is, therefore, a significant advantage, for multi-colour holograms in placing the master hologram (H1) in the exact position where it will inevitably be viewed as a complex real image when the finished (display or embossed) rainbow hologram is seen in white light.

In this way, the individual images for the colour separations can contain the relevant parallax information, in relation to the reconstructed images of the final hologram. But because we have spatially separated the linear recordings in accordance with their intended final wavelengths, the colours also appear in the calculated position.

In making an H1 hologram, conventional thought tends to make us believe that the reference beam has to be angled à la Upatnieks/Leith, with the object normal/perpendicular to the plate, but in practice, for a volume silver halide recording, there is absolutely no reason why the reference beam cannot be closer to normal ("head-on") whilst the object beam is oblique; the only inherent requirement is a suitable angle between the beams.

Of course, if the reference (dominant) illumination is approximately perpendicular to a glass plate, then an obvious problem is internal reflection ("wood grain"). Many holographers will spray the reverse (glass) side of the plate with black paint to be scraped off after processing; alternatively, a yellow- or red-dyed gelatin layer can be used to provide index-matched absorbance and may be peeled off or dissolved away in warm water. Alternatively, one can index match the plate onto a coloured glass plate of a colour complementary to the recording laser. The absence of the need for light to transmit through the plate when shooting is the advantage of transmission hologram recording, which allowed Agfa Gevaert to supply AH (anti-halation) coated plates in the early days of "Holotest" materials such as 8E56 and 8E75HD, and nowadays allows the inventive holographer to produce opaque plates by their own means.

So, in order to take advantage of Benton's observation that the transmission grating of a rainbow hologram will inevitably disperse collimated light of various wavelengths into a plane of foci which lies tilted in front of the hologram plate or film, there is no reason why we cannot place the master hologram in exactly the position that its real image will eventually be formed.

Recording the image components in the positions where they will finally be viewed helps to gain control of the imaging process and plan the final image. This situation differs from the Lippmann *reflection* hologram, whose wavelength filtration allows a more straightforward facsimile of reality. But the iridescent colours of the rainbow hologram have indisputably found a very special application in both display and security holography.

If we divide any object into colour zones, we can then represent these zones in separate or mixed colours in the final hologram. This colour separation can refer to graphic artwork or a three-dimensional model – the so-called "2D/3D" method described in Chapter 9 is a hybrid of these.

If there are three component colours, each will be recorded on the master plate on an individual linear slit whose position is selected in accordance with the required viewing colour (wavelength). *So we are, in fact, producing a final hologram image comprising a range of selected colours, by the use of only one laser.* In embossing origination studios, this laser will inevitably be blue, in order to allow the rainbow hologram to be recorded in photoresist as a surface-relief hologram that will provide a relief microstructure template for the creation of a nickel embossing shim.

However, for simplicity, if we consider for now the making of a hologram with a green laser, then the final position of the central view will be exactly as we set up the shoot on our optical table.

In Figure 8.40, the colour-separated artwork is placed in the oblique position shown in accordance with the calculation of the α-angle associated with the chosen reference angle of the final hologram. In the event that translucent, flat graphic artwork (photography) is used rather than a three-dimensional front-lit model, the illumination may be through a diffuser from the reverse side, and the artwork arranged on a glass surface perpendicular to the α-axis. In practice, film artwork can be positioned in register by using a "pin-bar". This would be the case in the event of stereographic origination of the type shown in Figure 8.39. Advanced techniques might equally use an electronic means of image projection on the table.

Using a single master plate so that all of the colour separations are recorded in register on that surface, with a single reference beam is, in the authors' opinion, a simpler way to ensure a precise result than the alternative methods of using multiple reference beams or using separate strips of glass to hold the three rainbow masters. Moreover, in this way, it is a relatively simple issue to record multiple wavelength components; thus, a yellow component of the graphic image could be included in the hologram *either* as a pair of exposures at the linear positions on the master hologram, corresponding to "red and green", or *alternatively*, at a position mid-way between these lines equivalent to a "yellow" wavelength of ~580 nm.

Whereas the simple registration is the advantage of the *single master, single reference* method for multi-colour rainbow hologram origination, this does introduce the problem that there is a need for a collimated H1 reference beam, as shown in Figure 8.40. In turn, this leads to a deficit of available light for the H2 transfer shown in Figure 8.41, when compared to the more traditional techniques associated with rainbow holography. However, another advantage of the single master method is that a single exposure is made to produce the H2, so that, in general, there is a distinct advantage in total

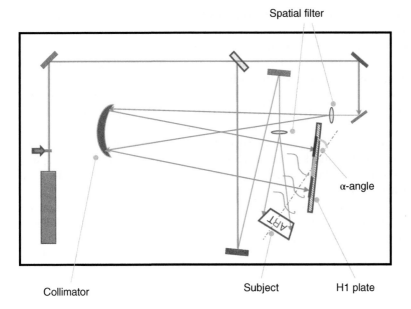

Figure 8.40 α-angle master hologram.

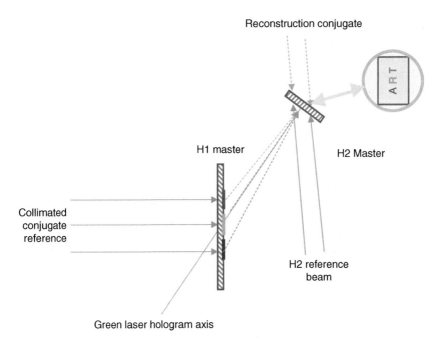

Figure 8.41 α-angle transfer to H2.

diffraction efficiency as compared to the methods where successive separate exposures are made on the H2 recording material.

On the single master plate, it is possible to arrange, in advance, the slit positions for the various colour component recordings, as shown with the designated "blue" zone close to the artwork. The green slit is naturally on the centre-line of the plate. The channels can be marked on the plate with black masking tape or overlaid on a black card mask. Animations and stereographic separations are positioned laterally on the master whilst the colour channels are longitudinal with respect to the axis of the reference beam, as shown later in Figure 8.43.

8.27 Producing an α-Angle H2 Transfer

The consequence of recording the master H1 in the format shown in Figure 8.40 becomes clear when we assemble the transfer camera to produce the H2 rainbow hologram. After processing the silver halide H1, it is returned to the plate holder illuminated by the collimated reference beam having been rotated by 180°.

Always keep the emulsion on the masked side at shooting so that the image is always displaced in the same direction when the plate is rotated. That way, the image plane will coincide with the routine position of the H2 plate in its plate holder on successive projects.

The outstanding weakness of the technique described is that the master hologram records the subject matter from different vertical perspectives for each colour. This means that the technique works perfectly only for planar objects. Many years ago I had a fascinating discussion with Nick Phillips as to whether the *stereographic* origination method would prove to be generally advantageous in embossed holography with regard to dispersion and blurring effects. In this instance, I believe there is a great advantage in the recording of stereograms in comparison with 2D/3D.

Where the object has significant depth, there are two approaches which present a possible solution. My limited opportunity to rotate the object with a precision turntable, whose axis coincides exactly with the exact image plane, appears to improve results. In the 2D/3D mode, offsetting the layers by trigonometry in accordance with their intended colour wavelengths and depth displacement is entirely successful.

The transfer of the real pseudoscopic image from the H1 into the H2 plate, as shown in Figure 8.41, produces a hologram which contains an image-planed graphic comprising three colour components.

Each of the fringe structure components will produce a real image of its source slit in a different position, as described by Benton, and each of those slits will pivot from the centre of the graphic image in the plane of the H1 master plate, as previously seen.

The image of each source slit is a rainbow, as predicted by the α-angle theory. As we see in Figure 8.42, the result is that the master channel, to which we have designated the "blue" artwork at recording, produces an image in the H2 which is dispersed into a spectrum whose central green band is raised above the normal, with the effect that the viewer experiences a blue image.

The "green" artwork (when mastering with a green laser) is coincident with the perpendicular; and the "red" artwork master channel produces a rainbow whose centre is below the normal, thus presenting a registered red view of its content in the plane of the

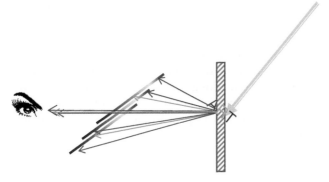

Figure 8.42 Alpha-angle reconstruction.

viewer's eyes when "red rotates radically" in white light illumination. Our viewer thus sees all of the colour components *simultaneously* in the correct focal plane *in the position that they were actually shot.*

8.28 Utilising the Full Gamut of Rainbow Colours

The dispersion of white light by a colour rainbow hologram produces a spectrum which is clearly not confined to discrete red, green and blue components. Once the geometrical calculations are made for the positions of the red, green and blue centre lines to be recorded in a transmission master, we have, in fact, the basis for a colour map of the components of the final transfer hologram, as mentioned above.

There is, then, no reason to limit the recording to three colour channels. As shown in Figure 8.43, the surface of an H1 transmission master, whether in the "parallel plates" or "α-angle" configuration, is, to all intents and purposes, a map of the colour and animation detail which will be seen by the viewer and, as such, provides a design template for the holograms made by the optical system. This is an invaluable aid to the graphic designer who will create artwork for future holograms from a set optical system. Therefore, this graphic designer does not need to have specialist knowledge of the holographic process; thus, we are en route to an efficient, professional hologram production infrastructure.

8.29 Reflection Hologram Transfers

From a full-aperture transmission hologram, it is possible to produce a reflection hologram transfer. Like the Denisyuk holograms discussed in Section 8.2, and whose existence was explained in Chapter 2, this is a Lippmann volume hologram whose planar fringes, predominantly parallel to the film surface, act as a reflective filter to the extent that a monochromatic image is seen when the hologram is illuminated in white light at the approximate angle of incidence of the original laser beam. For this reason, the reflection holograms we will make in this section will, unlike in the previous rainbow transmission section, essentially tend to be of the same colour as the laser we are using.

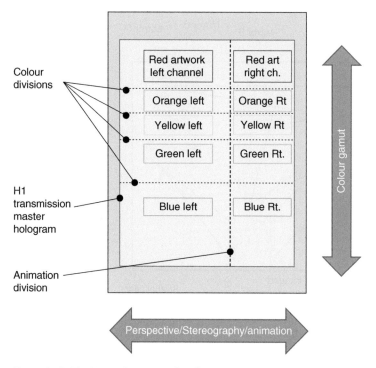

Figure 8.43 The H1 mask as a template for colour design.

In the case of silver halide holograms, it is relatively easy to induce deliberate fringe shrinkage in order to modify the colour to a shorter wavelength. These effects will be discussed later.

The image recorded in an H1 master, which is not slitted down as required for rainbow holography but recorded over the full surface of the first transmission master, as in Figure 8.28, can be transferred into reflection format, as shown in Figure 8.44. Here, the pseudoscopic real image from the transmission H1 is arranged to fall in the plane of the second generation recording plate. This is a similar situation as the previously discussed image transfer of a transmission hologram, but the master is not masked to a slit, because all of the parallax information is now required. Remember, this image is pseudoscopic, as if requiring to be viewed from the side where the master is situated. But, in practice, this is exactly what will happen, because the reference beam shown in the recording process will, in fact, be conjugated in the process of viewing the final hologram: it is as if that beam will come from the reverse direction to that shown in the transfer camera in Figure 8.44.

For this reason, when we describe shortly the process of examining the focussed image plane on a diffuser screen, you will soon come to terms with the idea of moving the H1 and H2 plates *apart* in order to make the image plane move forward.

In Figure 8.44, the laser beam is shuttered close to the output port of the laser. Such proximity can be very useful in terms of health and safety aspects of running a holographic studio, especially where high-power Class 4 lasers are in use, as mentioned in Chapter 7.

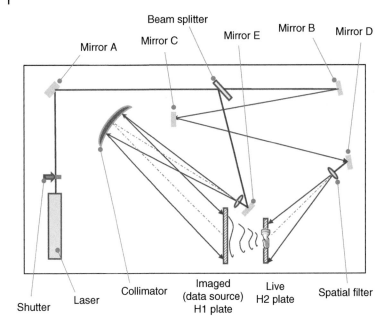

Figure 8.44 Reflection hologram H2 transfer camera.

Many of us who were introduced to holography in the early "helium–neon (Holographic Stone Age) era" were rather shocked to find that tried and tested shutters and light baffles were no longer appropriate for use with high-power argon lasers when embossed holography was introduced in the 1980s. The temptation to succumb to the universally requested, almost obligatory "Doctor Who" type demonstration of laser beams eerily passing through smoke-filled air, for visitors to the hologram studio, was not always entirely deliberate!

The corner mirror A and the beam splitter especially, and possibly subsequent optics in the beam paths, may receive high levels of power when Class 4 lasers are in use. Optics must be suitably selected specifically for their exact purpose when powerful lasers are in use.

Referring to Figure 8.44, the laser beam is divided at the variable beam splitter to provide suitable balance between the object and reference beams and, in this case, that balance may be assessed in the form of a beam ratio at the recording plate *of the order* of 1:1. The limitation of the amount of object light which you can permit varies with each image subject matter, but, in general, the point at which the object "burns out" the plate locally will dictate the shortest ratio, for the whole scene, that you can use to achieve the brightest possible hologram.

Of course, the H1 master is effectively focussing its 3D image on the surface of the H2 recording plate, so the intensity of the collimated beam of laser light on the H1 master in terms of mW/cm^2 is not necessarily difficult to achieve, provided the H1 master was made with high efficiency, which, as previously discussed, is prudently tempered by its requirement for a low noise level.

After setting the reference angle of the H2 hologram by the positioning of mirror D, the mirrors B and C can be moved to organise the path length equality to accommodate the coherence length of the laser. As described in Chapter 7, mirror E is deliberately placed as close as possible to the axis of the collimating mirror, since if this is of spherical profile, the emerging beam characteristics will improve as the angles of incidence and reflection reduce.

The H2 plate holder is used to hold a thin diffuser screen whilst the 3D image from the H1 is examined (with attention to safety) on the screen with a lens, or preferably a camera system to enable its desired plane of focus to fall on the surface of the recording plate, before the holder is finally locked down. In practice, the first hologram test will provide the best assessment of the image plane, since it is often a subject of choice regarding the effect when the hologram is viewed under various lighting conditions.

Once the H2 plate holder is locked down in its final position, it is then time to optimise the precise exposure conditions and, without fail, it is wise to decide upon a general exposure time before making a range of small adjustments of the beam ratio to optimise the hologram brightness and noise level. With experience of silver halide work, the holographer can often learn to judge these optimal conditions by examining the silver density of the unbleached plate in order to shortcut the optimisation process – the subject of the hologram will be clearly visible in the developer tray, but, if too clearly defined, the noise level will suffer.

We have found that, especially in the process of making reflection holograms, the joy and excitement of producing each new image has the ability to conceal a thousand sins! For this reason, there is very good cause indeed to keep your brightest and best hologram ever made continually at hand, so that you can make comparisons and temper your satisfaction with every new image *before* its release to public viewing.

Sadly, with this tip, you will often find yourself having to make the desperately difficult decision, at the H2 test stage, to go back to H1 recording and make a small improvement for the preservation of the highest standards.

Photopolymer reflection holograms have many similar characteristics to silver halide work at the H2 image optimisation stage. In this case, however, we have the remarkable situation where the image is visible during the exposure process (*still* remarkable after 28 years' experience of the material!).

As the image forms, we see interference between the new image forming on the film and the projected image from the H1 master. The two interact as phase change occurs and we see interference shadows running across the image zone. At the intermediate stages, the image zone becomes a remarkably efficient "interferometer" as regards deleterious motion caused by instability of any type. Tapping the table during this process will reveal a whole new spectacular aspect of motion-sensing interferometry as shadow fringes rush across the image area in dynamic style.

Photopolymer imaging of this type offers the amazing opportunity to block the original H1 source image "live" at the end of the recording process to reveal the existence of the almost identical image on the new film, in the absence of the original source data, simply by blocking the H1 illuminating beam. Enjoy!

8.30   "Pseudo-colour" Holograms

Multi-colour reflection holograms can be made by the pseudo-colour technique with the use of a single laser. The necessary chemistry is discussed in Chapter 6 and the high class work of Iñaki Beguiristain is famous for the colour effects and precision he is able

Figure 8.45 Pseudo-colour hologram.

to achieve. The hologram shown in Figure 8.45 was made with Andrew Rowe and Simon Brown in 1985 as a pseudo-colour transfer with helium–neon laser of masters made with pulse lasers at Applied Holographics [4].

Three first generation H1 masters were produced with a ruby pulse H1 system. The containers in the holograms contained indicator solutions which were adjusted between exposures by the addition of single drops of acid and alkali in order to adjust the transparency of their individual contents to the 694 nm ruby light. The advantage of ruby pulse exposure to avoid stability problems for the liquids was considerable, and the overhead reference system allowed the vessels to stand upright. The "yellow" stopper in the volumetric flask, which is just visible in the photograph, was actually red!

Simon Brown calculated the varying reference angles and beam divergence for the HeNe transfer and the Agfa 8E75 recording plate for the H2 was pre-swollen three times to achieve RGB colouration from successive 633 nm exposures before the plate was finally developed. We would shoot such "pseudo-colour" holograms in reverse order (Blue–Green–Red) in order to achieve similar exposure time for each exposure, with the effect that the more concentrated TEA was sequentially washed away from the film in the increasingly dilute pre-treatments. But, in this way, all of the exposures were subject to the *speed improvement* caused by the TEA treatment, despite the fact that the red exposure was not significantly colour shifted when the reagent was washed out.

Unfortunately, the chemical manipulation has taken its toll over the past 30 years and the hologram is hazier than it was originally and the gelatin appears microscopically cracked and damaged. In our experience, archival stability has otherwise been very reliable in silver halide materials whether they are hermetically sealed or not. However, on

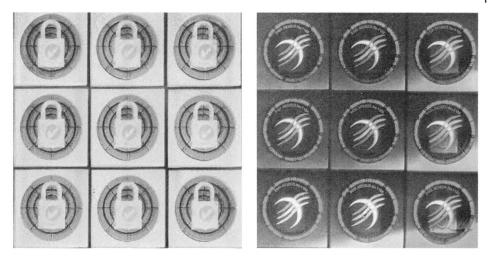

Figure 8.46 Separate channels of a full-colour reflection security hologram.

certain occasions images have faded, cracked or even disappeared. Dichromated gelatin holograms present a far greater challenge as regards ageing; complete hermetic sealing against air is vital.

8.31 Real-colour Holograms

Improvements in the recording materials available for reflection holography have made it possible to avoid many of the problems experienced in creating multiple colours in a single layer of a Lippmann hologram. Additionally, the advent of new sources of laser light in a range of colours makes it possible to produce RGB tri-stimulus images, with the advantages of stable colouration and a wide window of view. Jonathan Wiltshire produced the H2 contact master hologram seen in Figure 8.46 at Bowater Holographics [7] in 2015. This is a step and repeated glass H2 where vertical parallax is substituted with a secondary 2D graphic security channel.

The two photographs, taken at separate angles of declination, show a high degree of mutual extinction, with little ghosting, whilst retaining discrete and consistent colours which defy rainbow holography.

Notes

1 Saxby, G. (2003) *Practical Holography*, 3rd edition. CRC Press (ISBN: 978-0750309127). Holograms by Applied Holographics, courtesy of Opsec Security. www.Opsecsecuity.com
2 Kisak, P. F. (ed.) (2015) *The Gravitational Wave: Ripples in Space–time*. CreateSpace (ISBN: 978-1519665775).
3 Benton, S. (1982) "The Mathematical Optics of White Light Transmission Holograms". *International Symposium on Display Holography*, Lake Forest, vol. 1.

4 Holograms made by Applied Holographics, courtesy of Opsec Security.

5 McGrew, S. (1982) "A graphical method for calculating pseudo colour hologram geometries". *International Symposium on Display Holography*, Lake Forest, vol. 1.

6 St Cyr, S. (1984) "Holographer's Worksheet for the Benton Math". *Holosphere*, 12(8), 4–16.

7 Hologram by kind permission of Bowater Holographics. www.Bowaters.com

9

Sources of Holographic Imagery

9.1 The Methods for Incorporation of 3D Artwork into Holograms

In practice, there are limited techniques by which design concepts, artwork or subject matter generally can be introduced into holographic form. These are, in general terms:

1) The original concept of the use of holography to record the image of a real 3D object or model. With the aid of pulse lasers, unstable subjects and live subjects can be recorded. (Example: the VISA "dove".)
2) The conventional holographic recording of a solid assembly of planar graphics in the form of "2D/3D" arrays. The hybrid "object" or "model" is a stack of photographic glass plates or film bearing artwork in the form of text or line graphics. (Example: The Bank of England notes).
3) Digital holograms where the surface of the recording material is divided into pixels, tracks or voxels individually imaged with plane or complex gratings. The artwork tends automatically to be in the form of computer-generated graphics. These digital techniques were described in Chapter 4. (Example: The Euro banknotes.)
4) Stereographic methods of using photographic or computer-generated graphics to produce three-dimensional images by utilising the binocular vision of the viewer, without the aid of viewing devices traditionally associated with stereographic imaging such as the anaglyph method. (Example: The Shakespeare "Bard-card".)

My co-author makes an important point in his preface that, in the case of display holography, it is extremely important that holography keeps pace with the modern developments in graphics and interactive media. The digital and stereographic methods above will undoubtedly assume a special importance in future.

9.2 Making Holograms of Models and Real Objects

There is a basic requirement of a 1:1 size ratio for any model or object proposed as the subject of a hologram. This is a condition that can only be avoided by complex optical

The Hologram: Principles and Techniques, First Edition. Martin J. Richardson and John D. Wiltshire.
© 2018 John Wiley and Sons Ltd. Published 2018 by John Wiley & Sons Ltd.
Companion website: www.wiley.com/go/richardson/holograms

manipulation in the hologram recording process. In considering solid/real subject matter, we immediately encounter the requirements that:

1) Only subjects that are of the appropriate size are useful.
2) A model must be sculpted at the scale required. This frequently entails a specialist sculptor working with the aid of magnifying equipment and automatically, therefore, limits the level of available detail.
3) Polarisation is a very serious issue. A polarising surface to an object or model will always provide a crisp and clear image.
4) Multiple colour rendition of the subject is complex.
5) Supporting subject matter secretly is a special skill.
6) Q-switched pulse holography enables the use of unstable, moving and living subjects.

As discussed briefly in Chapter 8, a real object or model intended for holographic recording would ideally be lit by numerous light sources in exactly the same way that a photographic studio would expect to light a model, generally with dominant "key" light and important graduated "fill" lighting from other angles. There is no reason why this is not possible in holography.

The basic recording camera for transmission H1 holograms is described in Chapter 8. Subsidiary beams can be split from the object illumination beam to enable more elaborate illumination of the object. The idea of adjustable fibre optic delivery from the split ends of a bundle of fibres is attractive. In that case, it is to be remembered that, with limited coherence length, beam paths must be adjusted to accommodate the reduced speed of light in glass. By whatever means, the intention would be to create comprehensive lighting for the object with a specific key light and fill lighting to control shadow, whilst preserving sufficient 3D depth cues, which are, to a great extent, dependent upon tonality.

9.3   Models Designed for Multi-colour Rainbow Holograms

In rainbow hologram origination, there is a simple opportunity to introduce colour effects associated with each individual illumination source, and in particular, the obvious opportunity to use a separate colour for a backlit screen. We discussed in Chapter 8 how the vertical perspective of a 3D model changes in relation to the position of the individual colour slit positions for the H1 recording, so it is important to bear this in mind at the model design stage. For example, the perspective of a sphere does not change with angle, although lighting horizons might well do so.

9.4   Supporting the Model

An object for hologram recording must be of the exact size required of its image and it must be suspended in position *in a stable fashion*. This ability to support the object in a stable position with *invisible support* is one of the great skills to be attained by the holographer.

Three obvious categories spring to mind in terms of supporting the model:

1) The object stands on a solid ground with the opportunity for a deliberate shadow on the floor to enhance parallax. The ground plate is attached firmly to the holder for the recording plate.
2) The object is attached to a strong, rigid pin or rigid wire hidden behind its bulk. The wire is painted black, is not lit and is hidden from view. It must be attached firmly to the optical table, perhaps via a stand in shadow.
3) The object is mounted on a glass plate which might preferably be anti-reflection coated. The plate is firmly held in register in a three-point mount. This glass plate support method was used at Applied Holographics to enable the model to be removed, re-painted and replaced in exact register by Andrew Rowe and myself in the hologram in Figure 9.1 made by Applied Holographics for General Foods Inc. [1]. Here, the model was re-painted in white, black and grey between successive shots for a rainbow master. This was not achieved on the first iteration and numerous layers of paint were present before completion, to the extent that I was concerned by the model becoming incrementally larger! Fortunately, the requirement for the lettering on the bear to be white prevented any need to mask the script. An image designer must take such a feature into account to make the holographer's task achievable.

The holographic recording process offers a unique way to eliminate certain unwanted image components. It is remarkably frustrating to find that intended "black" backgrounds often image relatively brightly and, rather annoyingly, even incorporate undesirable shadows of the main subject matter. But where we have a requirement for an invisible backdrop to a scene, we can utilise an effect which is ordinarily one of our greatest enemies: movement, or motion.

Figure 9.1 3D colour hologram of model.

Figure 9.2 Fabergé egg clock.

As long as the backdrop in question is not associated with the stable support of the subject matter of the hologram, it is relatively easy to arrange for its total disappearance from the hologram recording. If the backdrop is moving at the time of exposure, it can produce the perfect "blackness" to contrast the bright image we require.

For example, one excellent method we devised for this purpose was to use vibration from a clock motor to provide a minute movement of a card screen during exposure, to ensure that the subject was not included in the hologram; interestingly, it does not matter whether that screen is black or white card (for visualisation of the scene, black material is possibly advantageous, and does not unnecessarily fog the recording plate with incoherent light). The effect of movement in the origination process is seen in Figure 9.2. Here, in a trial exposure depicting in realistic "true colour" the iconic Fabergé egg clock museum artefact of St Petersburg [2], the gold chain which hangs from the egg itself, below the clock face, being free to swing, has moved sufficiently during exposure to avoid imaging in the hologram and therefore appears as a "dense", black, three-dimensional shape, obscuring the surface behind it: a three-dimensional void which demonstrates strange full parallax against the remainder of the otherwise perfect image.

9.5   Pulse Laser Origination

The use of Q-switched pulse lasers eliminates almost completely all the concerns described above related to movement of the subject matter. The incredibly short duration of the pulse means that during the ~15 ns taken to create the standing wave, the object is effectively frozen in motion in exactly the same way that high-speed photography is able to capture an image from a moving subject in sport, for example.

In the case of pulse holography, the arithmetic is relatively simple, because the fringe spacing of a Bragg grating is of the order of half the wavelength of the laser light involved.

Therefore, for the use of a ruby laser to produce a volume grating:

$$\lambda = 2d \sin\theta$$

where θ = half the angle between incident laser beams. Thus, for two beams arriving from opposite directions:

$$\theta = 90°$$

$$\sin 90° = 1; \text{so } d = \lambda / 2$$

$$694 \text{ nm} / 2 = 347 \text{ nm}$$

At low levels of relative motion between the object and reference beams, the profile of the standing wave and the fringe contrast are reduced. If there is a net phase shift of one half of one wavelength during the exposure, the standing wave is entirely eliminated.

To get an approximate idea of the resilience of the pulse laser recording against movement effects, consider the movement of ½ wavelength during the exposure:

$$347 \text{ nm in } 15 \text{ ns exposure time} = 23 \text{ m} / \text{s} = 83 \text{ km} / \text{hr} \left(\sim 50 \text{ miles} / \text{hour} \right)$$

Rapid movement of this order will entirely prevent the recording of a hologram. Nevertheless, this type of limitation will still allow completely unstable subject matter, such as a moving human portrait, to proceed successfully.

Double-pulse interferometry has been used by groups such as Rolls Royce at Derby to analyse vibration in aircraft turbines and other mechanical systems. The working engine is exposed to a double pulse so as to record successive holograms which interact in the replay condition to show fringes of interference associated with air vibration. The time period between the two pulses is very short, and an optical system called a de-rotator is used to achieve realistic relative speed of movement for hologram recording.

It is to be remembered that a Q-switched pulse laser has both a "coherence length" and a "beam length". The coherence length of a JK Ruby pulse HLS10 laser is in excess of one metre, and will often record a hologram of a significant part of the whole studio. Looking into the extreme edges of a pulse H1 (as if looking into a window) can often reveal the working technician as well as the intended subject. The authors used such a laser to record images of a fruit bat in flight – the image is shown in Figure 9.3. The effect was to freeze the action of the bat in flight whilst detail such as the fur of the flying mammal was sharply defined. We understood that the flying action of a bat was quite different to that of a bird; the 3D action image is therefore a useful study aid. The dark room in very low level green light was an ideal environment for the nocturnal bat. At Applied Holographics, a laser trigger system had been built to allow the pulse to discharge at the correct moment in accordance with the interruption of a low-power laser beam. This equipment was not available to the authors when the bat hologram was recorded. Instead, the zoologist would release the bat to fly across the recording zone when the laser was charged and the deft timing of the trigger finger of the authors produced suitable masters with a minimum of film wastage!

Figure 9.3 **Fruit bat in flight.**

9.6 The "2D/3D" Technique

2D/3D has been one of the most influential techniques in embossed hologram technology. For the security industry in the 1980s, in the environment of the widespread growth of computer graphics technology, it provided the perfect means to harness holography and draw it away from its routine subject matter, which had, to a great extent, descended into the replication of miniature models derived from secondary sources such as the toy and jewellery industries. For example, my own earliest test holograms in the laser laboratory at 3 M Research in 1980 featured endless images of the then topical, now ubiquitous, *Star Wars* figures.

The basic concept of 2D/3D is the substitution of 3D sculptural models for the equivalent 3D array of graphics in layers which fulfil the hologram viewer's expectation for three-dimensional qualities, whilst introducing security graphics in the form of the text and guilloche-type line patterns which are traditionally associated with engraving and security print.

Typically, layers of artwork are assigned to glass lithographic plates, or films, which, when lit from behind, are then used as the subject for holograms in the form of a stack of plates with several layers of spatially registered coordinated graphics. The individual graphic plates can then be interchanged between positive and negative aspects of the artwork to allow views of the rear plates through "block-out" masks containing the foreground optics. In this way, we can produce layers of graphics in iridescent colours where the foreground images obscure those behind them with respect to the parallax available to the viewer, but the spaces between letters or lines still permit an unobscured "3D" view of the rear planes.

In Figure 9.4(a), the plate with graphic text "Foreground" is a clear plate with black lettering. It is important that the text is very dense in order to ensure no visibility of the

Figure 9.4 Principle of the "2D/3D" technique.

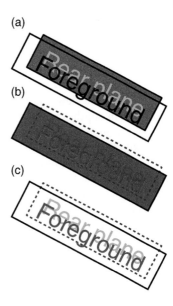

(a)

(b)

(c)

text "Rear plane" through the alphanumeric detail. It is equally important that the ground area is perfectly clear to allow the most translucent, non-scattering view of the alphanumeric "Rear plane". These requirements set a very difficult task for lithographic film and its processing – to optimise both maximum and minimum density of a film (D_{max} and D_{min}) in photographic processing is extremely demanding, as discussed in Chapter 6.

Now, the rear plate is conversely a dense black with its text perfectly clear. This is backed with a diffusing layer of glass or film and illuminated from the rear by the mastering laser. The text is recorded in the transmission master hologram with the foreground mask in front in the appropriate registered position, selectively blocking the illuminated rear plane in accordance with its own graphics.

In Figure 9.4(a), we see the block-out mask "Foreground" producing a shadow in front of the background image "Rear plane". This image is recorded in the transmission master in the "green slit" position by illuminating the rear of the assembled plates through a diffuser of glass or film.

The mask of the "Rear plane" artwork is then removed from the artwork plate holder and replaced with a *blank glass spacer* to retain spatial register, as shown in Figure 9.4(b). The graphic is shown in the diagram in grey, only for illustration purposes, on what is, in fact, a perfectly clear spacer plate. The front plane artwork plate "Foreground" is then placed in positional register in front of the rear blank spacer plate. Laser light (of the same wavelength as previously used) is then flooded through the plate from the rear through a diffuser and is recorded in the master hologram in the "red slit" position.

As explained in Chapter 8, when the master hologram is reconstructed with the conjugate collimated laser beam, the angle between each of the two slits nominally designated "red" and "green", with respect to the H2 (second generation) reference beam, determine that the resulting rainbow hologram will show the two planes of graphics in different colours, which may be organised to be red and green when the viewer holds the hologram in the intended viewing position, with the white light illuminating source also at the intended position.

As shown in Figure 9.4(c), the rear plane is seen displaced from the film surface in the final hologram, by the thickness of the glass plate mask and spacer, as green text, whilst the red foreground text typically appears in the surface of the film. That way, the foreground image appears sharp and clear even when the final hologram is viewed in rather unsatisfactory lighting such as diffuse fluorescent illumination. There is horizontal parallax, so that the viewer is able to determine clearly that the two image components are separated in depth, and this parallax is enhanced by the fact that the green image cannot be seen through the body of the individual letters of the foreground. Specialists such as De la Rue Holographics are able to achieve exceptional levels of precision in the spatial arrangement of various colour components and their related block-out masks in holograms of limited depth, to the extent where this *precision itself* represents a formidable forensic security device.

The technique of "2D/3D" shares with stereographic means of origination the difficulties of preparing photographic images on glass or black and white film, which have processing characteristics which can be very difficult to achieve, such as high contrast, high maximum densities and exceptionally clear layers. In addition, the use of film substrates, in place of glass, may introduce significant problems with birefringence, which difficulties tend to be relatively unexplored and rarely encountered when using conventional light sources, but can be catastrophically exacerbated in holography by our use of *plane polarised* lasers.

9.7   The Rationale Behind Holographic Stereograms

Following Lloyd Cross's famous demonstration of the inclusion of photographic images to create the "integral" hologram "The Kiss II" in 1974, it was clear that stereographic sequences of film provide a route to produce holograms, and, moreover, that *holography provides a supreme means to display autostereoscopic images.*

Prior to the development of holographic stereogram techniques, holographers were faced with the dilemma of finding, or arranging for the customised sculpture of, small models of the correct size for their hologram subject.

In fact, there began to exist a growing group of artists capable of producing precision models for the very purpose of holography, with the knowledge that, unless complex optical image-size reductions involving large-scale mirrors and lenses were involved in the mastering process, then the hologram image would be precisely the same size as the model itself. This led to significant difficulties, because the creation of miniature models for the holographic industry and, more particularly, following the success of the VISA hologram in 1984, the security industry, meant the industry would become reliant upon the skill of the individual model maker or sculptor and, furthermore, the models and, in fact, the sculptors themselves, would thus become a part of the security regime.

When using a physical model for hologram origination, there is much skill in supporting a delicate model in the desired position, with invisible support, whilst achieving the required positional stability for the mastering process, as described above.

In the case of colour rainbow holography, we discussed previously the difficulties of producing colour images from models in the single-laser techniques used for embossed holography, but, of course, the techniques of stereo-photographic data input for the holograms *alleviate both the scale and colour difficulties in one fell swoop.*

In Chapter 4, we briefly discussed the links which exist between certain forms of digital holography and the introduction of stereographic image data for that purpose, in order to create three-dimensional images.

Accepting these principles, we should look at the fundamental reasoning behind the link between stereographic photography and holographic origination techniques. To do this, it is instructive to consider the most prominent and routine means by which we make a typical transmission or reflection hologram and how that optical configuration relates immediately to the type of stereographic film sequences which can easily be recorded with an ordinary camera, or preferably more professionally with a purpose-built camera.

At Applied Holographics in the mid-80s, Andrew Rowe and I began exploring the possibilities of introducing 3D data from original stereographic recordings with the acquisition of simple, multi-lens stereographic 35 mm film cameras. Later, A.H. acquired A.D.D., with the effect that Craig Newswanger and Chris Outwater supplied to the A.H. unit at Braxted Park, Essex a superbly engineered, track-mounted, computer-controlled antique Mitchell cine camera capable of recording the highest quality film sequences in 35 mm 500 ASA Kodak colour cine film. Andrew and I hired the services of Jim Body, a very experienced cameraman from Pinewood, in order to improve our learning curve for cinematic techniques and we reached a suitable level of proficiency in camera work, lighting, film loading, etc. Jim's experience went back to the Technicolor era, where separate film strips were used to achieve colour cinema, and he was thus the consummate tutor for the demanding process of loading the Mitchell, which, to our chagrin, he considered a simple formality.

Applied's link with A.D.D. led me into many sessions of fascinating discussion and practical work with Craig, who is one of the most skilful and knowledgeable holographers in the world. These sessions took place in the UK and the USA, and we even spent an amazing week in Rome preparing equipment which had spontaneously developed an intermittent technical fault, in anticipation of a filming session involving His Holiness the Pope, prior to his state visit to Mexico. This film shoot never took place, but, after a nervous week of organising lighting equipment, attempting to solve an electronic problem in the camera control system and reconnaissance work in the Vatican, whose electrical wiring was, to say the least, ageing, the political discussion meetings continued through the week. The project was finally abandoned when I eventually heard on a hotel television news item that *even I* could comprehend with my very limited knowledge of Italian – "Il Papa è malato."

In order to understand the principles of converting stereographic film sequences into holographic form, it is instructive to consider the simple routine process upon which the production of a redundant transmission master hologram is based.

Imagine a classical H1 glass master hologram with its surface divided into a grid of small image zones, as shown in Figure 9.5.

The classical holography procedure to record this scene, described in Section 9.2, would normally involve illuminating the chess pieces in the diagram with laser light. Some of the reflected light then diffuses towards the recording plate where it interferes with coherent light from a collimated laser reference beam directed towards the plate, thus recording the interference standing wave as it exists in the plane of the ultra-fine-grain silver halide emulsion.

The recording of this interference pattern, after chemical processing, enables one to use the "H1" master – typically a silver halide volume transmission hologram – as a

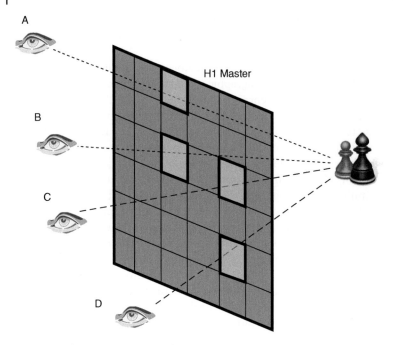

A

B

C

D

H1 Master

Figure 9.5 The stereogram principle.

window to reconstruct a three-dimensional image of the subject, by illuminating the plate subsequently with a conjugate reference beam. The moment when a glass master is replaced in the plate holder after chemical processing and illuminated with the original laser beam to reveal the new image never loses its thrill after a whole career in holography!

Thinking of this hologram as a window containing a three-dimensional view of the original subject matter, which retains all of the parallax of the scene recorded, we could literally divide the window into small, individual panes with strips of tape and consider exactly what a viewer's eye is able to see when looking through any of these limited areas.

Now, as these pixels reduce in size towards that of a pupil, each individual view of the subject through the master effectively loses its parallax. Each unit allows a single view of all of the subject, but there is no ability within that individual limited view to explore the parallax of the scene.

When my co-author invited Emmet Leith, at the turn of the twenty-first century, to visit T.H.I.S. at the Oxo Tower studio, I asked him whether he realised, prior to recording the world's first ever off-axis transmission hologram nearly 40 years previously, the full implications of the parallax of the scene.

He replied, *"I think you are forgetting the first hologram recording plate was only 1" square – there was no parallax!"*

Referring to Figure 9.5, an eye placed at position A sees the pawn obscure the bishop to a greater extent than an eye situated at position B. The eye at A also sees a view from above revealing the top of the chess pieces, but the eye at B sees a lower view. The eye at D sees the bishop obscure the pawn to a greater extent and has detail revealed under the base of the piece.

If each individual pixel reveals no parallax information about the subject matter, then, logically, the view seen through each small portion of the window can be replaced with a two-dimensional view (photograph) of the chess pieces with no loss to the fidelity of the scene *provided that each individual photographic view possesses the correct directional qualities and appropriate shadow.*

This simple phenomenon is the entire basis for the concept of the "holographic stereogram".

9.8   Various Configurations for Holographic Stereograms

Holographic stereograms, as we have previously mentioned, come in many forms, such as the semi-cylindrical type of Lloyd Cross, the embossed digital forms of McGrew, Haines and Toppan and the pixelated reflection holograms of Zebra and XYZ.

In the 1980s, A.D.D. produced "The Holodisk", which was presented as a film hologram disc on a rotating turntable lit from directly above. In the origination process, the film recording was from an elevated stationary movie camera and the subject was situated at the centre of a rotating turntable, with the ability to include animation in accordance with the speed of rotation.

During the mastering process, a large disc master was made which was segmented to allow the stereographic animated 2D cine film images to be transferred into the "polo-mint" master zone. When the final reflection hologram was made and illuminated from above, the miniature rotating turntable would display the contiguous animated stereographic images to the stationary viewer – as if viewing the original rotating turntable. Their eyes were in the zone where two coordinated stereographic views were always available to provide a 3D image. The size of the "polo-mint" master provided a realistic longitudinal window of view to the viewer looking down into the hologram and that aspect of the image coincided exactly with the original relationship between the recording camera tower and the subject matter, as shown in Figure 9.6.

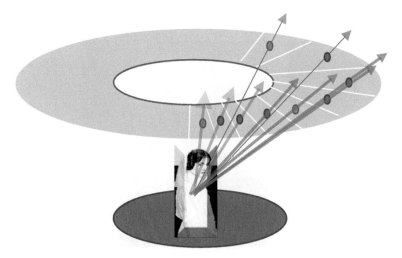

Figure 9.6 The "Holodisk" technique.

9.9   The Embossed Holographic Stereogram

The Shakespeare image applied to the cheque guarantee card, which became known as the "Bard-Card", sprung the Applied Holographics embossed security holograms to fame in 1988. Applied produced samples of this image in several sizes and there was some discussion as to the feasibility of an educational aspect of the project; whether a large-format hologram could perhaps appear in High Street banks to promote the security of the plastic card. Figure 9.7 shows two perspectives of an intermediate hologram which was produced as a 5" oval label as part of the proposed promotional campaign [3].

Craig Newswanger's stereographic cine recording of a Los Angles actor, a lookalike of William Shakespeare, was transferred by Nigel Abraham's team into photoresist, and as much as some of the British press criticised it (quote: "It's not even the *REAL* Shakespeare!"), the security industry loved it: everyone seemed to be astonished by the Bard's smile.

For many years, the image ran parallel to the VISA dove on plastic cards, but, unlike the many variable examples of counterfeit and compromise we witnessed to the Dove hologram, in no case did we witness any attempt to duplicate the image of Shakespeare. As a completely new technique it was, at that time, quite simply beyond compromise.

Following the basic principle described in Figure 9.5, the format of the first generation master allowed the image to be used in exactly the same way as a conventional hologram, but, unlike the full-aperture master shown in the diagram, the Bard image was a rainbow hologram where the colour information was able to be stored in separate lateral bands of the master (with respect to the upright image). So this configuration was the obvious route by which colour imagery of any chosen size could be introduced into embossed holography, and only later did Haines's computerised digital "one-step" method present a more direct method to show high-quality colour 3D images in embossed holography. The general sequence of operations in making an embossed stereogram provides a complex inter-related chain of events

Figure 9.7 Two views of the Shakespeare embossed stereogram.

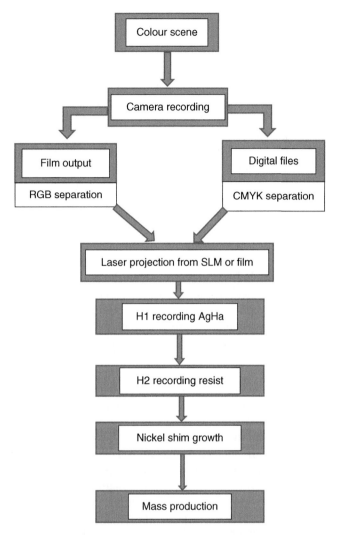

Figure 9.8 The process train for stereogram production.

shown in Figure 9.8; the security value is probably enhanced by the wide range of skills involved in a successful origination, because each individual step offers several opportunities for error.

Having spent much of his early career concentrating on pulse portraiture, my co-author is of the belief, with his extensive knowledge of the relationship between display holography and conventional art, that stereographic origination is now the most futuristic method of introducing data into holographic form, particularly in view of its advantages regarding incorporation of real colour, animation, freedom of scale and the advantageous substitution of laser exposure of live subjects for the more familiar photographic recordings. Stereo origination of holograms builds another bridge between holography and more conventional techniques in art and display.

9.10   Stereographic Film Recording Configuration

Of course, as well as introducing the advantage of an entry route into holography for new talent, the departure from pure holography towards this hybrid technique also requires the existing holographer to become proficient in a whole new area of photography. Nowadays, the great developments in computer manipulation of graphic images allow post-recording processing of images to correct anomalies which would otherwise prevent the production of first class holographic images. After more than two decades of study, these properties are now well understood, but at the time of the path-finding work at A.D.D. and Applied Holographics in the 1980s, the quality demanded by holograms for the banking sector meant that first class hologram production demanded first class photographic recording. The 35 mm footage from a movie camera was directly colour-separated in the lab and used for laser projection in the hologram recording.

In his 1991 paper, "Shear Lens Photography for Holographic Stereograms" [4], Bill Molteni neatly summarised the advantage of shear lens usage to stabilise the image position in a stereogram by pointing out that the important point of principle is to ensure that the camera recording and the holographic image recording on the optical table are both governed by similar geometrical layouts. He also makes the very important point that in terms of our stereographic perception, the provision of effective "depth cues" requires a great degree of precision in the position of image points in both their vertical and horizontal placement. Michael Halle's excellent thesis "The Generalised Holographic Stereogram" [5] also covered wide-ranging aspects of the technique in detailed fashion.

Molteni explained why the very worst camera recording configuration is perhaps the one that appears the simplest and thus attracts the attention of beginners in this field! That is, a camera which turns towards the subject whilst travelling along a straight track. This configuration introduces not only keystone distortion but magnification problems as the camera moves away from the subject. As explained by Rudolf Kingslake in his book *Optics in Photography* [6], lateral distortion is often acceptable in conventional photography, whereas vertical keystone distortion is usually highly undesirable.

A television presentation in the 1980s showed the science presenter Michael Rodd filmed with a camera on a curved track to produce a successful portrait. The distortions involved, however, were not flattering to the subject, and the geometry of the track itself meant that larger subjects than a human portrait were effectively precluded.

When all is said and done, our very precise purpose in creating colour holographic stereograms for embossed holography is to produce sequences of photography without vertical parallax in accordance with the rainbow principle, but which will stabilise the position of the 3D image of the subject in the film surface. This is eminently compatible with shear camera recording – the shear mechanism, where the lens and film are moving in parallel planes, will facilitate the master hologram being recorded onto a flat plate and will thus duplicate the viewing condition of the hologram. Then, the viewer's eyes are able to act with freedom to interpret the recorded scene, much as they would the depth and parallax of the original scene.

One of the most serious defects of an image-planed hologram, whether a stereogram or conventional recording, is where the image itself moves against the horizontal motion of the viewer's eye, frequently appearing to "fly out" of the film as the viewer moves away from the central view.

Assuming that the holographic optical table work as regards proper beam collimation etc. (described in Chapter 8) is followed, the shear lens method is the best solution for the avoidance of this phenomenon.

9.11   Shear Camera Recording

This method of recording, with the use of a wide-angle lens on a camera which does not turn towards the subject whilst moving along a track, will prevent many of the common image stability issues. The shear lens method also allows each component image to be recorded at an appropriate size on the film.

Because of the automatic re-centring effect, the moving lens places each image in the centre of the film frame. It thus allows far better image resolution, since the subject frame of interest fills the individual film frame so fully in comparison with the use of a simple, stationary wide-angle lens, which would record a *similar aspect* of the subject in *only a very small area* of the film. (One feature which never ceases to amaze when setting up a film shoot with the shear lens is the necessary width of the backdrop screen.)

In Figure 9.9, the subject is lit with the typical stationary illumination that we expect to find in a film studio: key light, "kicker" or a lit backdrop and a little diffuse flooding to control the level of shadow.

The camera runs steadily along a linear track, typically at, say, 0.5 metres per second or less. For the recording of life scenes, this allows plenty of time for the director to organise any animation of the subject. The shear lens is controlled by computer such that it moves relative to the film or digital light-sensitive array in the camera back. This lens movement effectively re-centres the image on the recording as the camera moves along its own track, and the incremental displacement of the active area of film from the lens, and the resulting expansion, ensures that the image size is consistent as the camera moves away from the subject.

When a human portrait subject is chosen for a hologram, the director of photography must prevent the sitter from following the camera with their eyes, as we are attempting to record the correct range of angular views of the sitter's face. Clearly, if they turn their head to follow the camera, then the stereographic information will be forfeited in favour

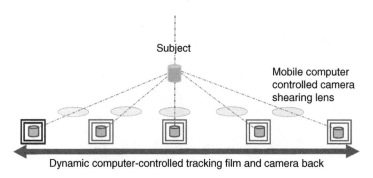

Figure 9.9   Shear camera mechanism.

of animation. The sitter should, therefore, fix their attention on a target such as a flagged lighting stand beyond the central area of the camera's motion.

There is a favourable balance of such perspective retention, against animation, which enables the hologram to achieve a good blend of solid three-dimensional appearance whilst including a realistic level of movement and interest in the final holographic image. Above all, the portrait subject must remain predominantly still during the recording of the central frames of film, and must not move forward or backwards in relation to the camera. The extreme views, as the camera traverses the last quarter of the track, may be allowed to contain a degree of motion; in that way, the sitter is likely to remain still in the critical central viewing zone, with the effect that a really solid and stable 3D image will be available to the viewer under any reasonable lighting conditions.

Due to the sensitivity of stereographic vision to height change, the track and the camera mounting we used at Applied Holographics were rigid and relatively heavily engineered. It is important to arrange the camera path to be level, and perpendicular in both planes, with the subject. As shown in Figure 9.10, the very stable design of the track would enable us to elevate it with tubular scaffolding to considerable heights. In Andrew Rowe's photograph, taken in the Church of Our Lady in Bruges, the author used a step ladder to work on the camera whilst the rail was elevated some two metres with scaffolding – he successfully produced a steady sequence of frames from the heavy Mitchell camera.

Figure 9.10 Working on the elevated Mitchell camera in Bruges.

Although the use of a tracking camera provides some difficult logistical issues with lighting equipment, unlike a rotating turntable arrangement, the linear camera track with shear lens allows perfect conditions for classical photo studio lighting of the subject. This static lighting provides ideal depth-cue shadowing for the three-dimensional medium. It is sometimes difficult to organise lighting stands without obstructing the camera path unless working in a well-equipped studio where lighting can be suspended from scaffolding or the roof structure.

It is my belief that, in view of the nature of holography, with so much special emphasis on the three-dimensional nature of the subject, and thus the temptation to exacerbate the depth cues in shadow detail, the use of relatively "hard" lighting is generally justified – a minimum of diffusion with lighting dominated by key lights. However, this technique does have the potential to offend the directors of photography or image consultants who frequently form part of the entourage of a leading celebrity!

In practice, any sequence of movie film where a camera moves steadily across a scene will produce good stereo effects. In large-format work, the precise details of the recording technique become arguably somewhat less important than the filming configuration for hand-held holograms of the type used for embossed security work. In fact, it is frequently the case when watching a scene in a movie that a holographer will sense a film sequence of camera motion which will automatically produce a first class stereographic effect.

In large-format work, the visualisation of stereographic and animation effects is largely at the behest of the viewer, via their restricted motion within the window of view of the hologram. The need to move in this way, and the reduced speed at which it can be achieved, appears to reduce the relative importance of the inherent stability of view of the holographic image itself, as related to its optical characteristics. Hand-held (security) holograms, on the other hand, tend to exacerbate any geometrical problems due to the ease of cycling the viewing process – the viewer will tend to tilt the hologram rapidly to and fro, to examine the animation and thus reveal any lack of continuity or unrealistic motion.

Whereas the possession of an elaborate computerised shear camera system of the type described is a serious investment, the involvement with alternative camera methods and software solutions is a "hit-and-miss" procedure. The best alternative "safe solution" is probably a rotating turntable for the subject matter, with a high-resolution camera to record the sequence of views. The axis of rotation for the table is then arranged to coincide with the important "image-planed" feature of the subject. For a portrait image, this would typically be the bisector of the eyes (effectively the bridge of the nose) of the sitter. This is shown in Figure 9.11.

The rotation of the turntable would then typically cover an angle representing the angle of view of the hologram. The motion of the turntable may be automated and a cine recording made. Alternatively, the table may be rotated manually, and still photographs recorded over a sequence of angular intervals; this method requires more cooperation by the model in the case of portrait work.

Unless the lighting can be incorporated on the turntable, the method has the disadvantage that shadows and highlights will move on the model with deleterious effect upon the viewer's recognition of depth cues, which are an important feature of three-dimensional perception.

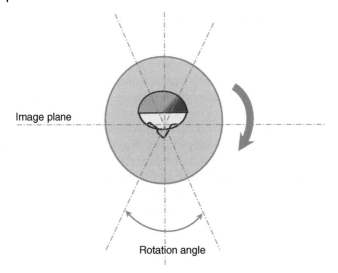

Figure 9.11 Centre of rotation for a portrait.

9.12 The Number of Image Channels for a Holographic Stereogram

The number of individual frames of photography used to create a stereogram is dependent upon several factors; obviously, the greater the number of channels, the greater the final continuity of vision for the viewer.

First, the practical reality is that the method of photographing the subject is a deciding factor in this issue, since a cine sequence will automatically provide a plethora of recorded data frames. A still camera with manual exposure will naturally tend to limit the availability of individual perspective recordings.

Second, the method of transferring the photographic information into a master hologram, in particular whether an automated method is available, will dictate how many individual image recordings can realistically be included in the first master, especially since colour work will require three sets of images for RGB.

Once the camera has recorded a sequence of colour photography with a suitable range of stereoscopic views of whatever subject matter has been chosen for holographic representation, that colour recording will need to be colour separated into separate tonal "black and white" sequences for inclusion into the holographic master hologram.

In the 35 mm film recording era, the film could be separated directly into black and white film sequences through RGB filters. These cine or still films were then suitable for use in laser-lit projection systems, or converted to transparency slides to introduce data sequences onto the mastering table by automatic or manual processes. The availability of such RGB *additive colour* technology was distinctly advantageous for holography as a means of acquiring films which allowed direct illumination by lasers to reveal the red, green and blue components of artwork. These would then be displayed in exactly that form in the final hologram, since holography also works in additive colour.

Modern technology means that the laser projection of an image can now be achieved in holography by the use of digital artwork to feed spatial light modulators such as LCD, LCOS or DMD (DLP) systems directly.

9.13 Process Colours and Holography – An Uncomfortable Partnership

Conventional printing technology has naturally directed its pre-press routines towards the use of CMYK *subtractive colour* systems as the print industry has been gradually developed, automated and digitised. It has therefore been necessary to adapt holography data import methods towards the acceptance and use of colour separations of the type universally made for the print industry.

It is not the intention here to go into the finer details of colour printing. However, it is important for the holographer to understand the relationship between the additive colour reproduction used in holography (and computer screens) and the subtractive colour system used in printing and fine art. Moreover, entry into commercial holography will involve regular interaction with graphic designers whose language is CMYK, and you, the holographer, have been appointed the official translator for the RGB linguists!

In order to have photographic or computer data transposed into the form of black and white film masks suitable for the holographic origination process, we have the choice of equipping ourselves with elaborate computer graphics systems for file preparation, PMT cameras with filters and image-setting facilities or using contractors for colour separation and image setting to supply black and white film greyscale separations for the optical table.

Traditional photographic separation of a colour photograph with red, green and blue filters will yield three colour separations which each contain the content of one of the primaries. When these primaries are re-mixed after transfer through a storage system, they are capable of reconstructing the original scene in a projector, video screen or *hologram.* But they cannot translate directly into printed matter, whose colour representation works by selectively *eliminating* the reflected light from white paper (*subtractive colour*).

It follows that a red-filtered separation, containing all of the red content of a colour image, will eliminate the blue and green components of the image when the subject is photographed in red light or with a red filter on a camera, as shown in Figure 9.12.

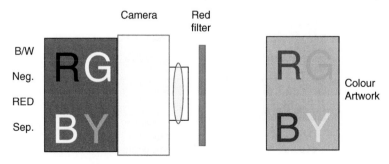

Figure 9.12 Separating a colour scene through a red filter.

The red filter will admit red components of each zone of the subject, shown on the camera back in the diagram as a negative image. But this could be recorded on either positive or negative film, or electronically. Light from the red letter is transmitted in its entirety by filter. Light from the blue and green zones of the subject is absorbed at the filter. Light from the orange and yellow zones is partially transmitted.

By inverting (creating a "negative" of) this tonal separation, which is, by definition, *devoid of the red information* that will later be printed by the mixing of Magenta and Yellow ink, we can produce the Cyan component of the printing process, as represented in Figure 9.13.

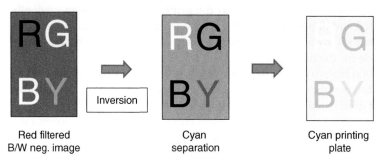

Figure 9.13 Creating the Cyan printing plate.

Likewise, in Figure 9.14, the inversion of the green separation, from which filtration has eliminated red and blue information, produces the Magenta component plate shown in Figure 9.15.

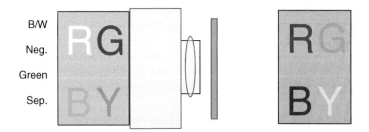

Figure 9.14 Separating a colour scene through a green filter.

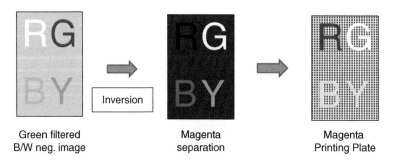

Figure 9.15 Creating the Magenta printing plate.

Finally, the blue negative, Figure 9.16, with red and green eliminated, produces the Yellow ink component shown in Figure 9.17.

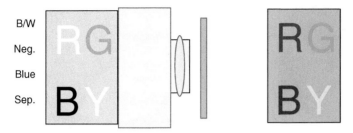

Figure 9.16 Separating a colour scene through a blue filter.

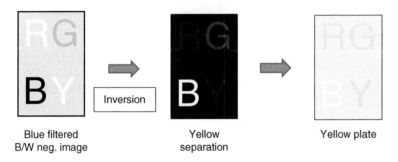

Figure 9.17 Creating the Yellow printing plate.

It is clear from the diagrams above, with their approximate tonal reversals, that if we use the *yellow separation* for laser projection, we will provide the *blue* image component for additive colour.

Now, as everyone who has experimented with watercolour painting will recognise, the effect of mixing cyan, magenta and yellow paint (or ink) is to achieve a rather distressing brown colour, so the print industry has compensated this practical failure by introducing a "key" plate (K), which is a black ink overlay to provide a richness to the dark zones. There may be cost advantages of reducing the quantity of coloured ink in favour of black.

The two colour models are summarised in the familiar colour diagrams shown in Figure 9.18.

The key to understanding the holography condition is the simple realisation that a holographic image begins with ethereal film propagating darkness – a black space – and conventional print technology begins, in almost all cases, with paper – a white diffuse reflector.

To realise our holographic image, we create gratings which diffract light towards the viewer. Without these gratings, she will see a *featureless black space*.

(a)

(b)

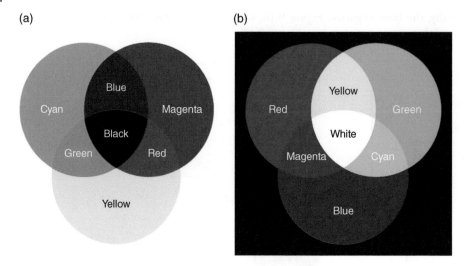

Figure 9.18 (a) Subtractive and (b) additive colour models.

To realise a printed picture, we obliterate the white paper with colour pigments or dyes. Without these pigments, the viewer will see a *featureless zone of white light.*

9.14   Assimilating CMYK Artwork with Holography

As described, in order to preserve valuable cyan, magenta and yellow colour ink resources, and to improve appearance and contrast potential in digital printing, exponents have designed a method where black ink is used to enhance the dynamic range of the process whilst reducing unnecessary usage of expensive pigments. Grey component replacement is one of the titles given to this type of adjustment. Ironically (or, as suggested above, predictably), this is so similar to the dilemma faced by the holographer.

Since our mastering process tends to introduce noise arbitrarily into the holographic image, particularly when blue lasers are involved in processing, there is a distinct advantage in ensuring that tonal black and white (lith) films used in the holographic mastering process for both stereographic and 2D/3D methods contain high-density black shadow in order to retain the required image contrast that we need in the final hologram. Every photon of laser light penetrating the image zones intended to represent darkness in the colour-separated films, or SLM projection, will play havoc with the desired conditions for a high-contrast, low-noise hologram.

To ensure that the best results are achieved, the black (K) separations supplied from an image-setter film process specialist need to be added to the density of each individual CMY film separation. There are several ways to achieve this:

1) Software manipulation to include the black component in the colour separations with the effect that the black density of each colour-separated film is increased. Printing houses are not usually familiar with this concept.

2) At the stage of hologram mastering by diffuse backlit illumination of colour separations in black and white film, we can place the black (K) separation of each image upon the diffuser screen on the registration pin bar. Then overlay each of the Cyan, Magenta and Yellow film separations in sequence in register on the same pin bar, so that, in turn, each of the component artworks will be presented to the mastering process with higher density and contrast qualities. This process can be problematic, especially with regard to birefringence and layer clarity, particularly in multi-plane 2D/3D holograms, where "foreground block-out" masks require extreme clarity.

3) Transfer the artwork from film above onto glass plates of suitable grain size and resolution. The processing of these plates must be planned carefully to achieve the highest layer clarity whilst simultaneously presenting high opacity or D_{max}, as previously described.

9.15 Interpretation of CMYK Separations in the RGB Format

Image-setting houses which are able to produce half-toned CMYK film separations are becoming less common as digital printing technology displaces conventional methods. When such transparencies are produced, the holographer uses the density enhancement methods above and can then convert the image detail directly into RGB in a way that is not, at first sight, totally obvious:

$$\mathbf{C}\text{yan} + \text{Blac}\mathbf{K}\text{ film separation pair} = \mathbf{R}\text{ed projection}$$
$$\mathbf{M}\text{agenta} + \text{Blac}\mathbf{K}\text{ film separation pair} = \mathbf{G}\text{reen projection}$$
$$\mathbf{Y}\text{ellow} + \text{Blac}\mathbf{K}\text{ film separation pair} = \mathbf{B}\text{lue projection}$$

By the 1902 example of colour printing in Figure 9.19, published by Dodd, Mead and Company [7], we can see the basic principle from which colour print has developed during the last century. Three-colour printing had been developed using cyan, magenta and yellow inks. The precise hue of these ink components is not necessarily tightly controlled, even today. The example shows the simplest three-colour print. Later, black ink was used to supplement darker areas to improve the appearance of the image as the technology developed.

There is a further issue of half-tone printing which should be taken into consideration when dealing with colour separations. Instead of the continuous tone or greyscales which photography offers, the image is divided into an array of dots. These dots are adjusted in size or spacing to allow higher or lower density to be achieved in accordance with the original photography or graphics. But the separation process will take into account problems which can occur such as the appearance of moiré patterns, which result from the interaction of the patterns of dots in the individual separations when they are recombined. Special ranges of angles are incorporated in each separation to avoid this issue. The holography process is capable of very high resolution transfer and therefore such moiré effects are prone to surviving into the final generation hologram.

Three color process

Figure 9.19 1902 colour print.

Notes

1 Colour rainbow hologram made for General Foods Inc. by Opsec Security Ltd and used with permission. © 1990 Post Consumer Brands LLC.

2 "OptoClone" egg hologram. Photo by M. Richardson; hologram courtesy of Alkis Lembessis, Hellenic Institute of Holography, Greece (HIH).

3 Hologram by Applied Holographics.

4 Molteni, W.J. Jr (1991) "Shear Lens Photography for Holographic Stereograms," *Proceedings of SPIE*, 1461, Practical Holography V, pp. 132–141.

5 Halle, M. (1991) "The Generalised Holographic Stereogram", S.M. thesis, Program in Media Arts and Sciences, Massachusetts Institute of Technology.

6 Kingslake, R. (1992) *Optics in Photography*. SPIE Publications (ISBN: 0-8194-0763-1).

7 Dodd, Mead and Company, 1902 (Wikipedia).

10

A Personal View of the History of Holography

The authors have been fortunate enough to have become involved at the embryonic stage of a young technology whose practical development has raced forward since the invention, in our lifetime, of the laser. This triggered rapid expansion which has captured the interest of both artists and technologists, and allowed devotees of those disciplines, such as the authors, to work together to achieve added benefits.

In this chapter we provide a timeline of events, many of which have involved us personally, that have caught our imagination as the *Age of Holography* has progressed.

1908 Gabriel Lippmann was awarded the Nobel Prize for his colour photographic recordings which relied upon interference recording in specially made silver halide plates. A layer of mercury was used to reflect light of different wavelengths back into the clear emulsion so as to create a standing wave, which was recorded in long exposures. These tended to be minutes in length. Reflective gratings were formed which enabled the reconstruction of the colour scene recorded by the camera. These images were difficult to view, but undoubtedly show the first recordings of natural colour due to diffraction.

Hans Bjelkhagen has studied this technique in great detail. This work is covered in Chapter 2 of his book *Ultra-Realistic Imaging* [1].

1947 Denis Gabor was a Hungarian refugee who fled the Nazis to become a British citizen. Working in England in electron microscopy, he is credited with the invention of the concept of holography. But in the absence of lasers, there was no real commercial value in this invention at the time – the resulting Nobel Prize did not follow until 1971.

1957 The race for the invention of the laser began in earnest. As described in Chapter 3 of this book, Gordon Gould and Charles Townes, who had previously invented the parallel concept (MASER) for microwave radiation, competed to produce the first demonstration of laser light. Gould was delayed by political problems as well as the difficulties of the technology, eventually allowing victory for Maiman at the Hughes Aircraft Corporation. This exciting race for laser light is brilliantly recounted by Jeff Hecht in his book *Beam – The Race to Make the Laser* [2].

1960 On May 16th, the invention of the ruby laser was demonstrated by Theodore (Ted) Maiman at Hughes. Soon, the resulting commercially produced ruby and helium–neon lasers would finally allow practical holography experimentation to proceed on a much wider scale.

The Hologram: Principles and Techniques, First Edition. Martin J. Richardson and John D. Wiltshire.
© 2018 John Wiley and Sons Ltd. Published 2018 by John Wiley & Sons Ltd.
Companion website: www.wiley.com/go/richardson/holograms

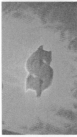

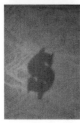

Figure 10.1 Benton rainbow hologram.

1962 Denisyuk in the USSR and Upatnieks and Leith in the USA "simultaneously" invented off-axis (modern) holography. Gabor's previous work with partially coherent mercury vapour lamps was limited by the on-axis configuration of the recording method. Now, the new techniques from East and West would allow a whole new imaging process to develop.

1967 A precedent was set at McDonnell Douglas (Conductron Corp.) when commercial ruby pulse lasers of the type invented by Maiman seven years before were used for the first time for portraiture.

1968 Stephen Benton invented the "Benton rainbow hologram," incidentally thereby enabling the functionality of the principle of surface-relief (embossed) holography.

Previously, transmission holograms would disperse white light illumination, but Benton's invention enabled iridescent transmission holograms to be viewable in white light for the first time; previously, a "blur" of white images was seen. He did not patent this brilliant invention!

Figure 10.1* shows the classical Benton rainbow effect where the 3D background line pattern changes through a range of colours, as proposed by Benton, as our viewing elevation (camera angle) changes, whilst the foreground logo, recorded in a second master slit, remains proportionately displaced in colour. The original, brilliant "single-slit" idea would thus eventually lead to sophisticated colour capabilities for embossed holography.

1969 Howard M. Smith (Kodak) wrote *Principles of Holography* – a predominantly mathematical account of a new imaging process [3]. It is now fascinating to review his mathematical explanations of our process, without reference to the practical developments which would later manifest themselves upon the process!

1970 Academics and artists were experimenting with holography to make three-dimensional display images for the first time as lasers became more freely available.

1974 Brilliant physicist Lloyd Cross made the first integral rainbow stereogram called "The Kiss" with Pam Brazier. This changed the course of display holography. The format was a semi-circular film display with integral lighting.

The indispensable book *Optics* by Eugene Hecht and Alfred Zajac was published [4].

1978 At the Royal Academy, HoloCo were receiving high-profile publicity for their display of holograms including the famous image of a water tap.

1979 John was working at 3M Research in photo emulsion research – experimenting with holography in an exploratory project with helium–neon and argon ion lasers. He had put in a record suggestion that holographic wallpaper could one day be used to make rooms appear larger ("Tardis effect") and thereby achieved authorisation from 3M management to familiarise himself with the technology as a supplementary project to research work for silver halide x-ray and graphic arts film products.

In that year, laser stage effects led the rock group The Who into an estranged marriage with holographic science. One key business investor, Keith Moon, was the first musician to be catapulted into the public eye as a hologram, in an exhibition created by the newly formed company HoloCo. The directors were the holographic scientist Nick Phillips, Senior Lecturer at Loughborough University, creative director Anton Furst, who was a film technician and lighting effects man, who later went on to design science fiction films such as *Batman* and John Wolff, who was involved with laser displays in the entertainment world. These pulsed laser transmissions show drummer Keith Moon drum-bashing in a holographic flurry of excitement, and, in retrospect, are historically important. Recently discovered material from that day offers a series of candid images depicting Phillips and Furst made shortly before the arrival of Moon. At the time, these could never have been thought of with any importance, but today they offer a candid window into a time of great excitement, optimism and a belief that, one day, holography would change the world! Interestingly, these were displayed at the Royal Academy in an exhibition called "Light Fantastic" and attracted huge crowds that stretched the length of The Mall at their peak.

This was something in which 1980s Madame Tussaud's took a keen interest, and they started to commission holographic tests, sponsoring Martin's master's degree at the Royal College of Art in London. Martin had the opportunity of recording both the wax casts of Queen Victoria and Marie Tussaud herself as experimental holograms and used the hologram as part of an installation at Victoria Station in London, where it remains to this day on the upper pavilion, central concourse.

1980 Graham Saxby wrote a practical guide *Holograms – How to Make and Display Them* [5]. This became the manual of operation for John's experiments in holography whilst at 3M Research. He was permitted to spend a fixed percentage of his working research time in the company's laser laboratory, and produced a number of transmission and reflection holograms on Agfa Holotest material. *Star Wars* characters were the main choice of subject. The holograms were lost in moving house in 2006 but John still likes to think he will find them one day, in searches of the dark and dingy loft.

John made high-resolution film coatings at 3M capable of recording holograms, based upon the existing Agfa and Kodak products, which were the only Western materials capable of recording holograms at that time. This 3M film still worked in 2012, so even if its imaging properties were not perfect, shelf life was good at least!

1981 The Light Fantastic Gallery in Covent Garden was exhibiting holograms commercially for the first time in London. John saved up for a week to buy John Kaufman's "Bird's Skull and Rocks" embossed hologram on one of his visits. 3M Research MD Alan Ferguson was partially sighted in one eye and observed that this hologram offered a form of 3D perception that exceeded the experience of stereo photography. Unlike stereo photography, the "real" parallax of holography allowed the viewer to move, to achieve a perception of parallax; one fully functional eye was capable of 3D perception. At this time, Nick Phillips at Loughborough University was developing a reputation as the leading British academic in holography. John invited Graham Saxby to speak as visiting lecturer at the 3M Research technical forum. When Graham mentioned future medical applications of holography, John asked whether x-ray lasers were feasible. The speaker responded that "in view of the power required to create coherent x-rays of suitable power, the question may be better addressed to a nuclear power expert, or possibly the US military as it is probably a key part of the rumoured forthcoming 'Star Wars' military initiative." Touché!

1982 Agfa Holotest materials at that time were capable of recording monochromatic reflection holograms of good quality. Covent Garden (Parallax Ltd) retailer Hamish Shearer recognised an opportunity for the lucrative sale of high-quality display images with co-director Larry Daniels.

1983 Hamish and Larry sold their patent to the newly formed Applied Holographics Ltd. On 4th September 1983, the *Sunday Times* featured an article by William Kay entitled "Hologram Snap Breakthrough" with Hamish, wearing safety goggles, operating a 35 mm film transport system which prototyped the HoloCopier.

The company had set up headquarters at Braxted Park in Essex, the estate of the late Michael Clark, former Vice-Chairman of Plessey.

John's colleague Reg Golder at 3M Research brought the newspaper article to his attention, and shortly afterwards Managing Director Alan Ferguson asked to see John and recommended that he applied to join the new holography company, bearing in mind his stated obsession with holography. As he sobbed uncontrollably, Alan assured him that 3M Research would miss him!

A laser-viewable Christmas-greeting transmission hologram was included in the application to Applied and John was offered the position of Holographic Chemist (3M colleagues taunted that you would only see him when light was shining from the correct direction).

On 10th October 1983, the authors attended the historic inaugural meeting of the Royal Photographic Society Holographic Group (see Figure 10.2) held at the Challoner Club, Pont Street, London and chaired by Michael Austin of the Central London Polytechnic. There was a definite air of excitement in this meeting as so many people dedicated to 3D imaging gathered together. The summer exhibition "Light Dimensions" at the RPS Headquarters in Bath had been a great success and had promoted widespread interest in holography. Michael Austin took the opportunity to check the number of left-handed people (rumoured to be exceptionally spatially aware [!]) in the room, and John remembers being smugly satisfied with the impressive result as one of the chosen few!

John was recruited to Applied Holographics in December 1983. The concept was to mass produce reflection holograms by the use of ruby pulse laser technology. After successful prototype work with Agfa Holotest 8E75 film in 35 mm sprocketed format, Applied's enquiry to Agfa for film production was received with limited enthusiasm, but AH received an approach from Ilford Ltd to produce a thin 63 μ polyester-base film on 9½" aerial spools. Applied immediately began technical discussion with Ilford regarding the design of improved film material for the stated ruby pulse application.

In 1983, Landis and Gyr invented the "Kinegram", the high profile of which endures today.

1984 AH commenced building the full-scale "Holocopier" in January 1984 in the style of the 35 mm machine that had captured the imagination of investors. In the era of monochromatic holography, the use of a ruby pulse laser to eliminate the enduring problems of movement ("motion" in US terminology) was a legitimate proposal for

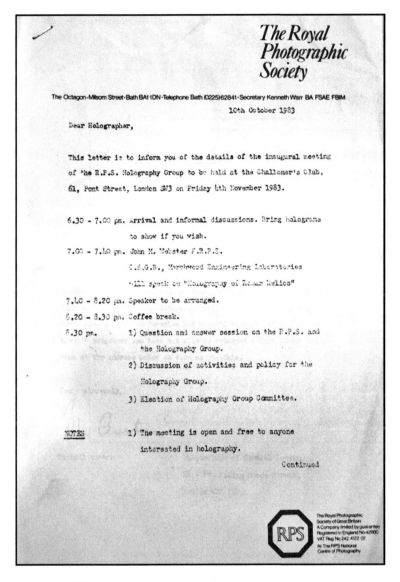

The Royal Photographic Society

The Octagon·Milsom Street·Bath BA1 1DN·Telephone Bath (0225)62841·Secretary Kenneth Warr BA FSAE FBIM

10th October 1983

Dear Holographer,

This letter is to inform you of the details of the inaugural meeting of the R.P.S. Holography Group to be held at the Challoner's Club, 61, Pont Street, London SW3 on Friday 4th November 1983.

6.30 - 7.00 pm. Arrival and informal discussions. Bring holograms to show if you wish.

7.00 - 7.40 pm. John M. Webster F.R.P.S.

C.E.G.B., Marchwood Engineering Laboratories will speak on "Holography of Roman Relics"

7.40 - 8.20 pm. Speaker to be arranged.

8.20 - 8.30 pm. Coffee break.

8.30 pm.   1) Question and answer session on the R.P.S. and the Holography Group.

                2) Discussion of activities and policy for the Holography Group.

                3) Election of Holography Group Committee.

NOTES   1) The meeting is open and free to anyone interested in holography.

Continued

The Royal Photographic
Society of Great Britain
A Company limited by guarantee
Registered in England No.42900
VAT Reg No.242 4122 02
At The RPS National
Centre of Photography

Figure 10.2 RPS inaugural meeting.

mass production of reflection holograms. But this principle did not meet with universal approval from other experts in holography.

National Geographic in March 1984 carried the famous Eagle hologram embossed in 11 million copies by A.B.N. with the article entitled "Lasers – A Splendid Light for Man's Use." It seemed that holography, as an industry, was taking off.

Applied Holographics was capitalised at £11m in July 1984 (Figure 10.3) on the Unlisted Securities Market (USM). The mass production system based upon 0.6J ruby pulse lasers from JK Lasers (later to become Lumonics) underwent intense R&D, with the late Rob Rattray, Dr Gavin Ross, ex-Ilford chemist Roy Trunley and John working with Farsound Engineering Ltd.

Copies of this document, having attached thereto the documents specified below, have been delivered to the Registrar of Companies for registration.

This document includes particulars given in compliance with the Regulations of the Council of The Stock Exchange for the purpose of giving information with regard to the Company. The Directors have taken all reasonable care to ensure that the facts stated herein are true and accurate in all material respects and that there are no other material facts the omission of which would make misleading any statement herein whether of fact or of opinion. All the Directors accept responsibility accordingly.

Application will be made to the Council of The Stock Exchange for the grant of permission to deal in the Ordinary Shares, issued and to be issued, and in the issued Warrants in the Unlisted Securities Market. It is emphasised that no application is being made for these securities to be admitted to listing.

APPLIED HOLOGRAPHICS P.L.C.

(Registered in England under the Companies Acts 1948 to 1981 with Number 1688482)

Placing

by

Laing & Cruickshank

incorporating McAnally, Montgomery & Co.

of

1,250,000 Ordinary Shares of 5p each at 180p per share

The shares being placed rank in full for all dividends hereafter declared or paid on the Ordinary Shares

---

SHARE CAPITAL

Before this placing:−

Authorised			Issued and Fully Paid	
£	No. of Shares		No. of Shares	£
50,000	1,000,000	Founders' Shares of 5p each	1,000,000	50,000.00
450,000	9,000,000	Ordinary Shares of 5p each	4,158,490	207,924.50
£500,000				£257,924.50

After this placing:−

£500,000	10,000,000	Ordinary Shares of 5p each	6,408,490	£320,424.50

Before and after this placing there are and will be in issue 1,141,550 Warrants entitling the holders to subscribe for a total of 1,141,550 Ordinary Shares at 30p per share. Particulars of the Warrants are set out in Appendix III.

INDEBTEDNESS

On 1st June, 1984 the Company had outstanding hire purchase and finance lease commitments of £45,386. Save as aforesaid, neither the Company nor its subsidiary had outstanding at that date any mortgages, charges or debentures, any loan capital (including term loans) outstanding or created but unissued, or any other borrowings or indebtedness in the nature of borrowing, including bank overdrafts and liabilities under acceptances (other than normal trade bills) or acceptance credits, hire purchase commitments or guarantees or other material contingent liabilities.

Figure 10.3 Applied Holographics placing.

With the help of Jeff Blyth, AH developed suitable chemical processing in the laboratory and converted this technology to machine-processing techniques rapidly with the expertise of Pat and Maurice Flaherty at Hope Industries.

The working mass production machine seen in Figure 10.4* was shown live on the BBC's science magazine show *Tomorrow's World* on 20th September of that year. This was one of the most terrifying experiences John had ever had; after rehearsal, at which we were castigated to the extent that the article might be cut, Rob Rattray and John made improvements to the presentation. We had been warned that as the title soundtrack to the live broadcast went up, we would feel some nerves! As Kieran Prendiville presented the article live on TV, John was on the floor inside the machine triggering a flash gun with red gel to simulate the pulse laser. *If only they knew...*

The BBC asked Applied to produce 2000 8″ × 6″ promotional holograms for free issue to viewers. The BBC workshop produced a model which was replicated as a Denisyuk recording in the HoloCopier. Elements of its structure were held in place with hot glue which melted during the first production run; thus, inadvertently creating one of the world's first holographic 3D movies on a 400′ roll of film!

The second run of production holograms included a mechanical enumeration device which made each of the individual holograms unique. Despite Rob Rattray's most intensive efforts, the electronic circuitry advancing the numbering device evidently took a secondary trigger from the 0.6 J pulse of laser light, so that a run of enumerated holograms with alternate numbers was produced; but suppliers of a suitably self-addressed envelope to Shepherd's Bush were not disappointed!

In November 1984, New Holographic Design, who had produced an excellent dispersion-compensated viewing device which facilitated a "black and white" image from laser transmission holograms, were working with AH to investigate possible mass production of the silver halide transmission holograms in the HoloCopier. Although the initial results were encouraging in terms of D.E., NHD had unfortunately chosen a film

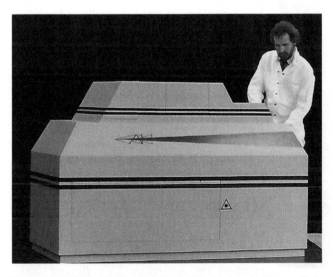

Figure 10.4 JDW with "The HoloCopier". Reproduced with permission of Dr Paul Dunn, Director of Optical Technologies, Opsec Ltd.

Figure 10.5 "Mirror Man" achromat.*

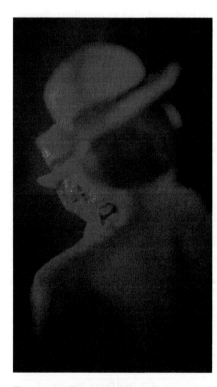

Figure 10.6 "Mirror Man" laser lit.*

format in excess of the 8″ web-width of the HoloCopier, so the relationship never came to fruition.

During 1984, the VISA "dove" embossed hologram began to appear on credit cards worldwide, bringing public attention to a new level.

In view of the work proceeding with New Holographic Design, whose viewing device permitted achromatic replay of transmission holograms, Adrian Lines brought a laser transmission version of his brilliant hologram "Mirror Man" to Braxted Park to experiment with transmission copying in the HoloCopier. The achromat in Figure 10.5 is shown as displayed with the NHD viewing device; in Figure 10.6 we see the original laser transmission hologram. The hologram is a classical demonstration of the parallax capability of holography. Adrian was tragically killed in a car crash at only 24 years of age. There was spine-tingling irony for those who knew Adrian, in that this young man's most famous hologram, a Denisyuk image of his own body presented in a coffin, was entitled "Portrait of the Artist as a Dead Man."

1985 Applied Holographics plc. undertook a first production run of 11,000,000 novelty holograms for Nabisco. The Ilford "HoloFilm" hologram was produced on 400′ rolls and, after lamination, was sheeted manually and applied by hand to printed cards. An abiding memory of that application process was that the real-time visual QC rejection of die-cut hologram sheets led to the operatives randomly depositing waste hologram sticky labels, to produce the unprecedented spectacle of a deep, *three-dimensional* pile of *three-dimensional* images. A close encounter with modern art!

In May 1985, the small research team at Braxted was also experimenting with machine-read holograms and HOEs for M.O.D. gun sights.

Further HoloCopier machines had been built for trials by major print companies. The machines were featured as an extensive presentation on Australian TV's *Beyond 2000* science magazine show.

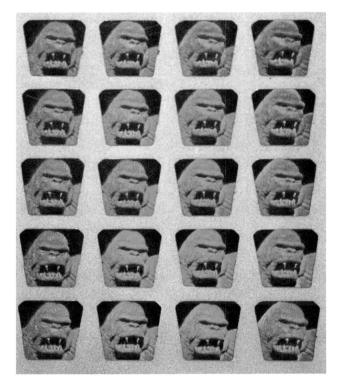

Figure 10.7 Die-cut sheet of Hasbro Visionaries holograms. Visionaries © 2016 Hasbro. Reproduced with permission.

Applied agreed with Canadian group NBS to form a new US Company, "Applied Holographics Inc.", in order to market products and technology across the Atlantic.

In July, with Hamish Shearer, John visited Jonathan Ross, Nigel Abraham and David Pizzanelli at See-3 Holograms. This was the first sign of interest by Applied Holographics in embossed holography. Jonathan, Nigel and David presented such a coherent view of the future possibilities for embossing that Hamish asked them to come to Braxted to have a look at the laser labs and to meet the company Chairman, Ossie Boxall, whose own previous experience with DataCard had raised his awareness of the immediate possibilities for embossed holography in the security industry.

National Geographic, in November 1985, carried the famous embossed skull hologram by A.B.N.H., "Search for Early Man".

Nick Phillips, Peter Miller, Graham Saxby and others were developing a strong holography unit at the Royal College of Art, where Martin was a student. Light Fantastic moved to The Trocadero with its exceptionally successful exhibition of Russian holograms based upon museum artefacts, "Treasures of the USSR".

1986 AH was running 100-million-unit runs of reflection holograms for Hasbro Inc. and Tonka in the USA and in England, having competed for the contract against the Polaroid photopolymer DMP128. The Hasbro "Visionaries" series (Figure 10.7) was a set of children's action figures, each knight having a chest plate featuring a monochromatic

Figure 10.8 Laser portrait of Simon Brown.*

reflection hologram of a fantasy character [6]. The photopolymer was able to provide extreme brightness of image for the toy company samples, but Applied's HoloCopier system had the ability to mass produce and die cut the required labels reliably in bulk at considerable speed with low waste margins. The bulk of production was conducted at Fort Wayne, Indiana, whilst masters were created in Essex. (Properly tuned ruby reflection masters appear so dim to the human eye that they are difficult to assess for QC purposes.)

Figure 10.7 shows a die-cut sheet of label product on black adhesive backing.*

In May 1986, AH inevitably began to take active involvement in embossed holograms. Dr David L. Greenaway joined the Board of Directors of Applied Holographics. David was formerly with RCA and Landis and Gyr, being involved with diffractive machine-read systems such as the telephone card, and brought his specialist knowledge of shim-making into the company, simultaneously with an agreement with Transfer Print Foils Inc. to form a foil-manufacturing facility (TAPF) in Washington, Tyne and Wear. Nigel Abraham also joined Applied to bring his own considerable expertise in embossed holography into the company.

David Greenaway then proceeded to set up a pre-production facility for shim origination in Switzerland under the name Applied Holographics Embossed Ltd.

The Braxted unit acquired a 10J ruby pulse laser capable of human portraiture. Having set the recording camera in operation with elaborate variable dielectric beam split and beam profile monitoring facilities, built by Ken Rickwood and David Winterbottom at the Essex University Industrial Physics department, after completion of energy calculations, AH commenced recording portrait holograms, and Figure 10.8* shows the first ever portrait recorded at Braxted, for which project leader Dr Simon Brown volunteered to be the first subject; a slightly cautious first pose where our incomplete confidence in our energy calculations led Simon to cover one eye in the first shot! Before inviting further sitters for our portrait camera, we took professional advice from Bryan Tozer at his newly formed LaserMet company [7].

Applied was working with Rolls Royce in non-destructive testing techniques for aircraft turbines, and missile air-flow visualisation for A.W.E. to produce white-light-visible copies of interferometric engineering analyses.

Index-matched scanning HoloCopiers were built in England and the USA by Applied. John and Craig Newswanger had recognised the demand for full-colour reflection holograms. Intense research work at Braxted by Simon Brown, Andrew Rowe and John led to success with active index-matching solutions, but the efforts to patent this technique were confounded by an astonishing similarity with prior art from Stephen Benton and Julie Walker [8].

1987 AH made a presentation to the US Bureau of Engraving regarding holographic protection of the dollar bill. Dr Paul Dunn joined AH from Third Dimension. His former employers attempted to prevent the transfer of his talents initially, which led to a temporary injunction. Paul has a very extensive knowledge of the physics of holography and was able to combine these scientific skills with the ability to organise the extremely complicated time schedules associated with the logistics of running several origination laboratories, enabling time pressure to be relieved from the labs with good effect. The addition of established holography experts was rapidly building a very strong team at Braxted Park.

1988 In January, Du Pont [9] scientists Bill Smothers and Dalen Keys worked at AH during their visit to England, on the early versions of the Du Pont holographic photopolymer, later to be called Izon*.

The "Denisyuk" holograms pictured in Figures 10.9 and 10.10 were made during the trial work at Braxted Park.* Due to the "live" recording process which occurs in photopolymer work, the closer interrogation of the watch hologram reveals a fascinating effect which is in strict contrast with the exposure properties of silver halide

Figure 10.9 Animated watch.*

holography. The stepping second hand of John's watch was imaged in several discrete positions during the first seconds of the exposure process. At that time, the photopolymer was flooded with white light and baked after laser exposure to produce a bright, monochromatic image. Use of the kitchen oven to bake the first odorous samples led John into a significant dispute with the catering department!

In *The Sunday Times* on 3rd April 1988, Hugh Pearman reported an interesting new project for Applied Holographics. John went to the race circuit at Swindon to fix individually enumerated reflection holograms to car parts in the Vauxhall Lotus Challenge competition. This was a race format where the principle of testing the drivers required that the cars and key engine parts were identical. The film holograms were affixed with permanent adhesives to ensure that all key components of the vehicle were standard.

In August 1988, *New Scientist* reported Paul Hubel and Andrew Ward's work at Oxford in real-colour holography. Hubel

Figure 10.10 Photopolymer coin hologram.*

Figure 10.11 Craig and Daniel approach the Vatican.

related the gamut of colour available to the hologram to the use of a wider range of laser wavelengths, working on Ilford recording materials.

The "Bard Card" (Shakespeare) embossed hologram by Applied was released to APACS for a new credit card. The project was reported positively in the *Sunday Times* Innovation supplement on 18th September 1988.

The public response led AH to transfer the main focus of its attention to embossed holography. Applied conducted a share issue to enable the transfer of production facilities to Washington, Tyne and Wear and a link with TPF which would be called Transfer All-Purpose Foils (TAPF).

Stevie Nicks was filmed at AH Inc. for an embossed hologram "The Other Side of the Mirror," which was used for album promotion in the UK.

National Geographic carried the embossed pulse hologram of a globe.

Graham Saxby was awarded an Honorary Fellowship of the Royal Photographic Society as a result of his work in Holography and Photography. He had by then retired from his position with Wolverhampton Polytechnic to become Senior Lecturer at the Royal College of Art.

Martin (now Professor of Modern Holography at De Montfort) achieved Britain's first PhD in holography from R.C.A. and began working at Darkroom Eight in Acton, London.

In November 1988, the Royal Photographic Society produced *The Photographic Journal* in a special issue relating to holography with articles by Margaret Benyon, Graham Saxby and Martin. Applied produced a pulse-laser-originated reflection hologram of a siskin (a type of bird) for the cover.

1990 On 10th January 1990, John travelled to Rome to work with Craig Newswanger to attempt to film Pope John Paul II in advance of his state visit to Mexico.

The subsequent project report to Chairman Ossie Boxall was entitled "How We (Almost) Shot the Pope". Applied's newly constructed computer-controlled stereogram rail with an antique Mitchell 35 mm cine camera was transported by air to Rome.

With Chris Outwater, Craig and John transported the camera system to a studio in Rome's Cinecitta, which is a studio regarded as Italy's answer to Hollywood. There, they were able to set up the system working in advance of the anticipated move to the Vatican, whilst Craig and Chris joined Dan Lieberman of Hologramas de Mexico in meetings with Cardinals and Mexican officials in an attempt to bring the project to fruition. Craig and Dan are seen heading towards the Vatican in Figure 10.11. As a result of the project, John was fortunate enough to receive a guided tour of St. Peter's, with the opportunity to visit the awesome "Wall G" in the crypt with the remains of St. Peter.

Figure 10.12 Pope John Paul II in 2D/3D.*

Political restrictions and, finally, the Pope's illness prevented filming. Holograms were eventually made from official still photography for the Mexican Government. We were relieved to find that, whereas the routine robe worn by the Pope was white, on certain occasions he was attired in red, and this was the photograph used to create a stereographic "2D/3D" image of the Pope by separating components of the image such as the Papal ferula and arms, the facial plane and the rear plane featuring the Vatican City. A three-dimensional stereographic cloudy sky was used as a backdrop (Figure 10.12).

The first dot-matrix holograms were made at Applied.* Andrew Rowe and John received components from Dotz! inventor Craig Newswanger and began to set up a working origination table at Braxted Park based upon a low-power HeCd laser. The first sample was low resolution (shown in Figure 10.13), which fortuitously provided a remarkable three-dimensional effect! At that time, Landis and Gyr and CSIRO were promoting their digital holography technology. But the intention at AH was to provide a particularly economic route toward digitisation.

Other holographic images made during that era included a holographic portrait stereogram that was never released. At AH Inc., Craig filmed stereogram footage of Michael Jackson for embossed hologram labels for perfume and for large-format holograms. There was a licensing problem, which meant that this exciting work was never seen.

1991 Canadian entrepreneur Dr Bill McGowan hired AH to undertake a tour of Europe to film Michelangelo sculptures for large-format stereographic display holograms.

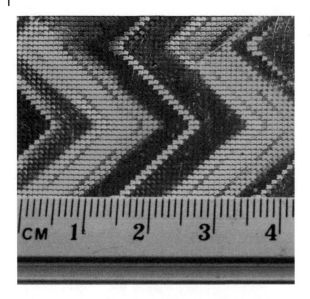

Figure 10.13 Braxted's first dot-matrix hologram.*

JDW set off with Andrew Rowe with the stereo camera in a hire truck. The first port of call was the Church of Our Lady in Bruges, where they met Bill McGowan, who had achieved access to the sculpture "Madonna and Child" in the church whilst it was closed to the public for the lunch period. They were to vacate the building in three hours. After completing the recording, there was a moment of amusement where John's inexperience and lack of confidence in the use of a film-changing bag to can the exposed film safely led him to seek an area of subdued light. He found that the empty confessional was ideal, but when he dropped the lid of the film can in the darkness there followed an expletive which is rarely heard in the church environment! At completion, Bill took the exposed film can to send to Craig Newswanger at AH Inc. for processing and gave us cash to get to Rome through Belgium and France. Figure 10.14 shows John returning to the van after a strenuous interrogation at a border post!

Driving through the Mont Blanc tunnel and the snow-covered foothills of the Alps, the equipment arrived in Rome to set up the camera track in three locations to record the Michelangelo works of Moses at San Pietro in Vincoli, Christ with the Cross at Santa Maria sopra Minerva and finally, La Pieta at St Peter's Basilica. In Figure 10.15, the horned Moses oversees Andrew and Bill McGowan's arrangement of lighting for his sculpture.

Returning to the hotel on the second day, the hire van developed a problem – the clutch cable broke – and Andrew achieved hero status by driving the van with its irreplaceable cargo on a hair-raising ride across central Rome without the ability to stop – or, at least, if he had stopped we would not have been able to start again. It fell to John to call Paul Dunn in the early hours of the morning to ask for help with getting repairs to the van and Paul astonishingly managed to use his exceptional organising skills *during the night* to convince the van hire company to get an engineer to the van within a few hours so that the filming stayed on schedule. Figure 10.16 shows Andrew assisted by the incredibly enthusiastic and helpful technicians at the Vatican, unloading the camera gear. Figure 10.17 shows La Pieta with colour card and slate prior to filming.

Figure 10.14 The journey to Rome.

Figure 10.15 Andrew with Bill McGowan.

The broken clutch episode was not the last driving drama of the tour. On the return journey, they were scheduled to record Michelangelo's wooden model of the Façade of San Lorenzo in Florence, but when John's fine navigation led them into a pedestrian piazza, the van was confronted by an armed policeman. Andrew had the presence of

Figure 10.16 Andrew unloading cameras at the Vatican.

Figure 10.17 Michelangelo's La Pieta.

mind to announce the camera equipment as "BBC" and we were warmly allowed to continue! I anticipate that "Brexit" is unlikely to enhance this type of cooperation.

Back in the lab at Braxted, John made a demonstration rainbow hologram from a smectic LCD panel. This LCD presented each pixel in a scattering or translucent mode. This was shown to Stephen Benton during his visit to AH and he revealed that MIT had already made similar holograms.

Andrew Rowe and John began work on the development of full-colour embossed holograms at AH with a view to creating 2D/3D and stereographic master images for embossing, by use of Benton's α-angle method. They used both film from stereographic recordings and 2D colour photography separations in the 2D/3D style. One of the first successes with this technique provided a three-dimensional colour image of the Microsoft logo by the use of laser-cut metallic lettering in the company font.

Glaxo then used the full-colour Union Flag hologram John produced with Andrew to protect the drug Zantac with a tamper-evident label. The new drug had suffered a counterfeiting scandal in Greece. Glaxo challenged AH to produce a new genre hologram in full colour and our research in colour rainbow work paid full dividend.

The Microsoft package Windows 3, released in May 1990, with an embossed stereogram portrait, was counterfeited in China. Initial despair at Applied regarding the counterfeiting of the stereogram produced by AH Inc. turned to celebration when the poor-quality fake hologram was deemed responsible for the actual discovery of the fraud. We were informed that the packaging, data discs and handbook were indistinguishable from the genuine Microsoft package, but the milky embossed hologram had allowed warehouse supervisors to spot false packages. Later, it was said that the hologram counterfeiters were apprehended with a small fine.

Hans Bjelkhagen and the pulse laser group Holicon Corporation set a new precedent in pulse holography with the recording of President Ronald Reagan at Santa Barbara, California [1].

1992 The Museum of Holography in New York closed for the last time after 15 years of existence.

Wembley Stadium began to protect its tickets with a holographic foil stamped in the shape of the "Twin Towers" (Figure 10.18). This was very satisfying, for the company to present a solution to the long-standing problem of ticket forgery. A generic hologram was tested for the specialised hot-stamp and later completed with a customised image. The era of F.A. Cup Final attendees on TV in the morning of their club's big day, with tears in their eyes as they lamented the purchase of fake tickets, was finally behind us!

1993 At Braxted Park, with Jonathan Ross, Patrick Boyd created stereo film sequences for his remarkable full-colour rainbow holographic stereogram "Virtual Dialogues." Patrick was filmed dressed in operating theatre attire, with his "new born" baby with prosthetic umbilical cord, whilst the projected backdrop represented the child's VR experience.

IHMA was formed [10]. This organisation has since demonstrated its support for the practitioners of legitimate holograms and its role is now, of course, further necessitated by the modern proliferation of hologram counterfeiting.

Hans Bjelkhagen published his book *Silver-Halide Recording Materials: for Holography and Their Processing* [11]. John was introduced to this book many years later, by MR, and is quoted, "Incredibly, this publication did not come to my attention for many years

Figure 10.18 Wembley Stadium hologram.

following publication. I believe that many younger people find it difficult to believe that the internet did not become mainstream at all until the late eighties. Even then, the early search engines were incredibly laborious to use – I recall being introduced to Google at De la Rue Holographics by Simon Cooper around the millennium, with absolute shock at the ease of use. Computing came upon us very suddenly! After my own college experience in the late 60s of producing Fortran punch cards for computer-controlled experiments, Amstrad PCs launched in 1984. I think my younger son Darren, born in 1983, was perhaps one of the first generation of children to be able to write his name on a keyboard before he could write it on paper!"

In 1993, Applied was concentrating on optical and mechanical step-and-repeat "combination" methods to enable the inclusion of twin techniques in embossed hologram shim masters to provide increased security levels for leading-edge security holograms. The combination of techniques provides a significant barrier to counterfeiting. For example, stereographic 3D images can be combined with dot-matrix technology. A prospective counterfeiter is thus required to be au fait with both types of image as well as mastering the actual combination techniques. The Christmas card hologram (Figure 10.19)* made by Claire Lambert using an optical combination of the "laser foil" and "dot matrix" was an excellent demonstration of the registration of the two components of the image.

1994 John was developing LCD image projection systems with David Winterbottom [12] with a view to the automation of the stereographic origination system at Applied Holographics.

The Applied Holographics stereo camera track system was transported to Acton, West London, by Rob Rattray and John to film a stereo sequence for Martin's hologram "Rocker" featuring a rider astride a Harley-Davidson motorcycle. This full-colour rainbow hologram was transferred into resist by Craig Newswanger for an embossed hologram. A second image of a female rider was also produced, as shown in Figure 10.20. Later, MR allowed Mike Medora and Nigel Robiette at Colour Holographic to produce a 50 x 60 cm display hologram of the image of the male rider – this large-format hologram was known as "Biker". When this hologram was later displayed, on a dark evening, in the window of Martin's T.H.I.S. studio at Oxo Tower, London, JDW had the experience of standing behind a family group viewing the hologram. Father gave an eloquent explanation to his family of how the "projector behind the screen" was projecting sequences of perspective views such that the children were able to perceive a three-dimensional animated image...*if only they knew!*

The New Zealand stamp that AH made with Southern Colour Print, commemorating the 25th anniversary of the moon landing, was

Figure 10.19 Christmas hologram.

Figure 10.20 Biker stereogram.

Figure 10.21 The end of the road – that sinking feeling!

released on 20th July 1994. Applied's hologram stamp of H.M. Queen Elizabeth II was also released in the Isle of Man.

Applied supplied a large dot-matrix hologram to Pepsi for the two-litre bottles, the specific design of which, in a single, large, animated graphic provided a demanding task for the origination department, as described in Chapter 4.

1995 Martin opened his new company The Holographic Image Studio (T.H.I.S.).

At Braxted Park, Essex, Applied Holographics' automatic stereogram laser origination rig was completed and used commercially for the first time, based upon a Sony liquid crystal display tablet of the type that was designed with the original intention of replacing the hand-drawn slides for overhead projector display. The use of a high-resolution colour TFT screen with a single laser wavelength represented a relatively inefficient use of the pixel array, but results exceeded the quality of the commercially available alternative black and white screens in terms of resolution and contrast ratio.

One of the most amusing events in Braxted history occurred when a lorry driver came into the building to hand over a parcel without taking into account the value of a good handbrake on the precarious sloping car park outside the AH reception area (Figure 10.21).

With a rapidly assembled human chain, staff helped the desperate driver unload the parcels, as water from Capability Brown's lake slowly filled the truck! The driver phoned headquarters to assure them that the submerged parcels were rapidly drying out without damage! *If only they knew...*

Anaït Stephens produced pseudoscopic sculptures at Lake Forest and, working with Hans Bjelkhagen, produced colour holograms by direct "Denisyuk" illumination of the models to produce colour, real-image holograms.

Figure 10.22 The end of the road – that drinking feeling!

1996 Applied Holographics received a team of technologists from INCM Lisbon, led by Alexandre Cabral and Ana Andrade from INETI, for training in holographic techniques, prior to INCM's involvement in embossed holography facilities in Portugal.

1997 John travelled to Lisbon to install a holography origination system at INCM Lisbon which included an LCD stereogram origination projector. On a single table, the system was able to produce 2D/3D and stereographic images and transfer silver halide masters into photoresist surface-relief holograms.

After 14 years' service at AH, John eventually left the company. This was an emotional event and he is pictured in Figure 10.22 surrounded by gifts from my colleagues, including a not insignificant quantity of fine wine!

As a contractor, he later worked for AOT on the development of the hologram-protected Windows 2000 discs for Microsoft, developing shim-etching systems to personalise the graphics on generic data discs to differentiate various versions, language groups etc. He also returned to Lisbon to improve the performance of the multi-function origination system as an independent consultant.

1998 Martin recorded a hologram of the film director Martin Scorsese when he visited London to promote his movie *Casino*. Scorsese talked at length about 3D and later Martin invited him to become part of a project he was working on to capture the creative 1990s "Hidden images of Holography".

The authors later recorded pulse portrait holograms of Will Self and Billie Piper at T.H.I.S. in the Oxo Tower building on London's South Bank.

The loss of an embossing shim during transport at Orly Airport caused a serious setback in the design of the forthcoming Euro series of notes. The hologram design was themed in the style of the note print itself, and led to speculation about the need for complete re-design and consequent delays. In May 1998, the *Guardian* pointed out the irony that the shim was actually of no value whatsoever to the thieves, as it was clear that the design would be cancelled if stolen prior to the initial print run.

Will Self wrote an article in the *Sunday Times* magazine on 3rd October 1998 called "Will Self Confronts his New Alter Ego" about the experience of pulse portraiture at Martin's T.H.I.S. studio in his own inimitable style in his regular column.

A new digital technique for display holography was being explored at Zebra Inc., who created a promotional hologram of Ford's P2000 Prodigy concept car, in which ten tessellated hologram tiles made up one very large, computer-generated, colour reflection hologram. This was sensationally unveiled by none other than Bill Clinton, the President of the United States, at the winter 1998 Detroit Motor Show.

1999 Martin received an invitation to visit David Bowie in his working studio in New York. It was a surreal experience to meet such a celebrity in what appeared to be a decrepit warehouse in Varick Street, Manhattan. The penthouse suite was a hive of activity, however, with the Chung King Studios receptionist opening a sliding glass door beside the desk to reveal a long corridor; at the end stood the figure of David Jones, a.k.a. Bowie. The blond-haired superstar took time out from his Eidos project, "The Nomad Soul" with co-producer musician Reeves, to view MR's full-colour holograms. Martin describes a "star-struck" feeling of awe as he found himself presenting his art to one of the greatest innovators of the century. Suitably impressed with the concept of colour 3D imaging, Bowie invited Martin to bring the stereogram recording equipment to film him in London on 18th June. The stylish concept set for the album *Hours* is pictured in Figure 10.23 prior to the star's arrival. Martin directed as we filmed 20 minutes of stereogram sequences at Big Sky Studios, London. One hundred and three feet of Kodak Vision 550 film was recorded, as shown in Figure 10.24, and, on this occasion, the phrase "check the gate" held a very special significance (and even more so since David's death).

Always the true gentleman, Bowie chatted with us after the shoot, and when asked to autograph John's son's guitar, he coolly asked, "What shall I write?" Despite prior warning, from Martin, of the New York "heart in mouth" experience, John was lost for words for the first time in 46 years! The words "We Can Be Heroes" echoed round and round inside the star-struck head, reached the tip of the tongue, echoed round the head, tip of the tongue...silence. David wrote "**Hi Jon, Rock on. David Bowie 99**".
Martin later produced 1 metre square format display lenticular images and smaller versions which were included in Martin's book *spaceBomb* [13].

At Oxo Tower, the authors recorded pulse masters of a large 3D detailed model of a building proposal. John Perry transferred the 2 × 1 metre large-format landscape of the Chelsea Library development with spectacular results: a rainbow hologram with a truly awesome 3 metres of image depth.

Emmett Leith, Hans Bjelkhagen and Graham Saxby visited the studio at the Oxo Tower in London Blackfriars.

Figure 10.23 The set awaits David Bowie.

2000 John was now working at De la Rue Holographics with Brian Holmes, who has a truly remarkable technical understanding of the practical and theoretical aspects of security holography. The Bank of England Darwin £10 note hologram was produced. DLRH specialise in conventional holography 2D/3D techniques with graphics registration at very high levels of precision; Brian's ability to get his department to achieve absolutely precise image resolution between colour components sets the holograms apart as a leading security device.

2002 Euro notes in seven denominations began to circulate with holographic protection. The Bank of England £5 (Elizabeth Fry) note featuring the DLRH hologram, in similar style to the precision 2D/3D interleaved animated images of the previously issued £10 and £20 notes, was released. Producing security holograms of this quality entails weeks of fastidious attention to detail in the design studio and well-equipped holography laboratories at DLRH.

2003 On 9th November, Stephen Benton, inventor of the rainbow hologram, died. His legacy was effectively the whole of the embossed holography industry, now rated in billions of dollars annually.

Counterfeiting of embossed holograms including the Euro notes began to present a growing problem. I discussed with DLRH whether it would be prudent to consider involvement in volume reflection holography, but the company was reluctant to consider diluting concentration on the company's thriving embossing facilities.

August 10th was the hottest day ever recorded in England. It also happened to be the day chosen for Colour Holographic to record the film footage for the Nike hologram campaign featuring the prominent French rugby union scrum half Frédéric Michalak. The venue was the 3 Mills Studio in East London and in advance of the approaching Rugby World Cup, Michalak was flown into London for the shoot. Bearing in mind the

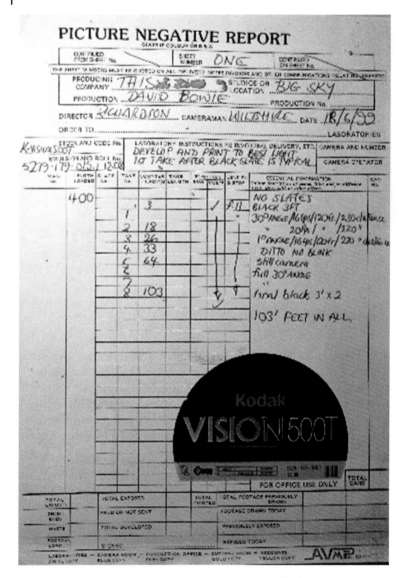

Figure 10.24 Picture neg. report.

need for a series of highly animated 50 x 60 cm action portraits of the rugby star, Mike Medora and Nigel Robiette had arranged a large stage with the vast backdrop required for shear camera recording. Of course, the consequence of illuminating such a vast scene for moving subject matter is the need for copious lighting! Nigel had arranged 130 kW of lighting to accompany the record-breaking ambient temperature. The result was that the control system and power supply for the camera track were at an unprecedented temperature, and they did not react favourably! Despite losing part of the angular function of the camera, the calm directorship of Robiette and Medora was able to modify the subject motion to accommodate the limitations with great success.

The holograms produced after processing and significant retouching with respect to the requirement for the ball to "fly out" of the hologram in some cases, were highly successful as an advertising campaign. Colour Holographic posted a video of the display in the Champs-Élysées [14], which attracted significant public attention both to Nike products and to the efficacy of holography in the advertising market.

2004 Smart Holograms Ltd was formed in Cambridge after research by Chris Lowe's department with assistance from Jeff Blyth.

Ver-tec Security Systems Ltd was formed with an initial contract with Merck Chemicals to explore aspects of full-colour reflection holography, and our first task was to summarise and define for Merck the technical value of the Smart Holograms project.

Figure 10.25 Biometrigram.

At Ver-tec, John drafted and filed five patents in reflection holography. Ver-tec produced and patented the prototype "Biometrigram" system (Figure 10.25). This was a table-top system which included a proprietary fingerprint scanner to feed an SLM projection system. A 10 mW helium–neon laser was used to image silver halide film produced by Colour Holographic. A self-contained desktop chemical processing unit then produced a two-channel reflection hologram on silver halide film. The process, issuing a hologram in "real time," could then be verified against a new, live scan of the fingerprint or against recorded data.

2005 One of the inventors of off-axis holography, Emmett Leith, died.

Rudolf van Renesse published a third edition of his book *Optical Document Security* [15], which has extensive explanations and assessments of security technology, including the various forms of holography.

John was working as a consultant to Smart Holograms to demonstrate mastering techniques for reflection holography. Ver-tec produced more patents for reflection holography. SHL decided to use a rotary exposure technique with embossing shims as masters for a drum replicator, producing reflection holograms which were to be used as moisture sensors. The silver halide film emulsion was presented emulsion-out in the final label, with the effect that humidity was able to influence the image colour of the reflection hologram label. Smart discussed this technique as a possible monitor for contamination of aviation fuel. Another possible security application involved the inclusion of a subjectively "invisible" image component which was revealed in an interactive step by the simple act of breathing upon the label: with the reputation of gelatin as a culture medium for micro-organisms, this innovation had initially been undermined in the environment of the SARS outbreak which threatened a pandemic!

2006 Yuri Denisyuk, perhaps the world's most famous holographer, died.

Denisyuk's legacy, the *single-beam hologram*, means that his name is literally on everyone's lips and has today presented a new challenge to replicate reality!

Bundesdrückerei published passport documents with personalised photopolymer reflection holograms.

Bayer Materials Science invented a new panchromatic photopolymer for holography, "Bayfol HX" [16].

Merck contracted TD&I Ltd to produce demonstrators for full-colour reflection holograms and discuss their relationship with sensor holograms. John assisted Merck with IP for sensor holograms.

In 2006, Ver-tec Systems purchased TD&I Ltd from David Winterbottom.

2007 Patrick Flynn, Satyamoorthy Kabilan and Jonathan Wiltshire joined Ver-tec; they created the HD3D* hologram and produced the full-colour animated stereographic Rubik's Cube demonstration hologram shown in Figure 10.26. When Satyamoorthy Kabilan presented these at a conference in the USA in 2008, several major players asked for samples, but the 35 mm production technique precluded genuine commercial interest. The sequence of images photographed shows, through the camera angle range, the steady "real" colour of the cube. As the hologram viewer moves left and right, its central block rotates and subtle "flick-in" images appear fleetingly at the extremes of tilting the hologram.

Martin was awarded the Saxby Medal by the Royal Photographic Society and appointed Professor of Modern Holography at De Montfort University, Leicester.

2008 On 13th March, Martin took part in an historic project involving Mike Medora and Jeff Blyth at the National Maritime Museum, Greenwich, which set another milestone in the concept of using holography for the archival presentation of museum artefacts: in this case, the famous 1759 timepiece "H4", designed by John Harrison, which had changed the world of navigation in the eighteenth century. At last, "H4" would become an "H1" in hologram parlance!

Figure 10.26 Ver-tec's Rubik's Cube.

Jeff had deliberately selected this particular evening to expedite his idea, just prior to the arrival of British Summertime, because it was necessary to take advantage of the early evening darkness for the recordings. Without the luxury of proper darkroom facilities, the light-sensitive holographic BBV recording plates supplied by a remarkable holographer, Mike Medora, would be subject to unusual levels of ambient light. That evening, Mike had travelled from Trinity Buoy Wharf, local home of his company Colour Holographic Ltd, with the recording plates. The holographic camera frame Jeff Blyth had designed and built was unloaded from Martin's car. The holographers met Mr. Jonathan Betts, Senior Specialist in Horology at the National Maritime Museum, who had greeted the project idea with enthusiasm, for he is constantly looking for new and exciting ways to interpret and reveal collections for visitors. Betts is a global authority on the timekeeper, having penned a number of authoritative works on the subject.

Knowing the recording equipment would take some hours to set in position and test, the group prepared for a long night. Waiting at the entrance of the observatory on this stormy night, there was delight to see a fantastic laser beam illuminating the Meridian emitted from the top of the observatory – fantastic indeed, as it lit up individual rain

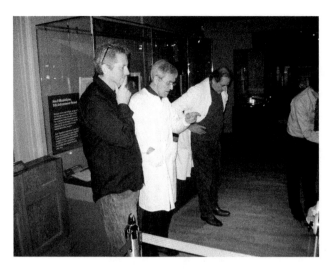

Figure 10.27 Holographers in pensive mood.

droplets which passed into its path, making each one sparkle like highly polished emeralds; treasure falling from the sky. The green laser is known as *The Prime Meridian Laser* and had been installed eight years earlier to mark the turning of the Millennium. The beam represents a projection of the line of "0" degrees longitude from the top of the Royal Observatory and highlights a stunning night-time view of London from Greenwich Park. The thought crossed Martin's mind that the use of this laser to record the hologram would be the definitive coup! On further investigation, however, the meridian argon laser beam turned out not to be of suitable beam quality.

The laser beam can be seen from as far as Pole Hill in Chingford and officially forty miles north. To point a thin line of laser light into the night sky – into the outer atmosphere – into deep space without the loss of intensity before moving into the path of an asteroid, perhaps once charted by the astronomers of Greenwich, sparks a strange sense of primeval freedom and childlike wonderment.

Inside the Royal Observatory, the holographers were led to the Horology Conservation Workshop (Figure 10.27) where they set up the holographic recording system and put into place the more suitable, but tiny, d.p.s.s. laser on loan from Klastech GmbH [17], which pushes out the impressive power of 150 mW.

Directing this laser beam toward the Blyth frame, Martin began to expand the laser beam and arrange a spatial filter, while Mike Medora set up the processing area using trusted chemistry – pre-mixed solutions contained in recycled plastic pint milk cartons! Jeff Blyth took great care to reassure Jonathan Betts, and his colleague David Rooney, as to the non-hazardous nature of the process to follow.

At 11 p.m., after some hours of preparation, the holographers were ready to make the first recording. They were led from the workshop across the cobbled courtyard that Newton once trod to "The Longitude Gallery" where "H4" is on display. They marched single file, through a security check and were counted one by one, then paraded crocodile fashion through CCT Security.

There, "H4" was positioned inside an extremely large, purpose-built toughened glass case. Betts, complete with mauve-coloured protective gloves (Figure 10.28), slid open

Figure 10.28 Jonathan Betts prepares Harrison's "H4".

the case and Martin reported goosebumps appearing on the back of his neck. "H4" was released, and placed into its unique titanium transport carrying case. Betts carefully placed "H4" onto the specified position in the holographic camera and, after checking laser power, light coverage and that this was not a dream, the holographers were about to make history in a way Harrison could not have anticipated in that building a quarter of a millennium previously. Mike Medora removed an unexposed glass photosensitive plate from its safe box and, with the assistance of Jonathan Betts, gently positioned it above "H4" to rest just above its surface. The ever-resourceful Jeff Blyth made final adjustments to the stability of the system, using razor blades to dampen any vibration, like a waiter would a rocking table. A 20-second exposure was estimated. The room itself was the very room where the world's "state-of-the-art" timekeeping trials were carried out, so historically there were leading-edge scientific resonances with the room itself! Selecting one of the louder "ticks" echoing around the workshop, the holographers started the whispered countdown…three…two…one. Martin raised the makeshift shutter, blocking the laser light and muttered the words, "*To the spirit of Harrison*". In silence, the holographers observed the beautiful green laser light illuminate this marvel of timekeeping mechanics. The hologram was recorded as they watched; the timekeeper's glass reflecting an eerie light to brighten the room as early morning rain tapped gently on the window. How strange this scene would have appeared to the outside world: windows flashing with otherworldly light, a bone-chilling storm, perhaps some uncanny experiment was taking place inside conducted by "mad scientists"?

If only they knew…

Mike processed the plate to reveal the first successful archival recording of the Harrison watch that changed the history of navigation. His relief at the success of the in-situ holography studio is clear in the triumphal photograph (Figure 10.29)!

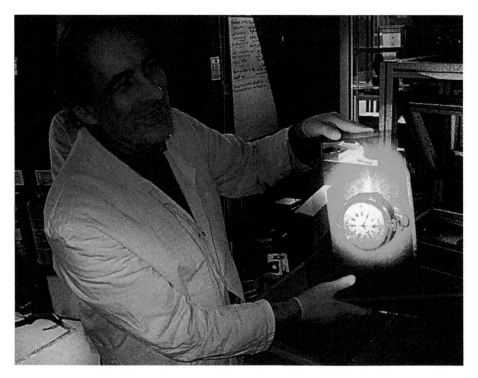

Figure 10.29 Mike with the successful "H4" hologram.

2009 On 23rd May, Professor Nick Phillips died after a long illness; Nick had been inspirational at Loughborough and RCA, and his broad knowledge of so many aspects of holography was unsurpassed. Nick had inspired so many people to take an interest in lasers and holography by his association with The Who and Britain's first really high-profile exhibition at the Royal Academy.

2010 John formed a company, 3DOS, with David Winterbottom, Patrick Flynn and Richard Evans. His son Jonathan Wiltshire began to make masters and samples for customers and 3DOS worked with BMS and Colour Holographic materials to produce reflection holograms on silver halide and photopolymer for security work.

2011 Step-and-repeat systems for master holograms for photopolymer and silver halide contact masters were successfully commissioned and new lasers were acquired to power the mastering systems. Demonstration edge-lit holograms were made with high efficiency for a car manufacturer for lighting systems.

2012 John later formed AOSYS Ltd, with Jonathan and with Patrick and Anthony Flynn.

2013 AOSYS was acquired by Bowater Holographics. BHL immediately expressed the intention to become involved in mass production as well as the hologram mastering technology that had been devised at AOSYS.

2014 Kurz's new "Kinegram Volume*" reflection hologram method was first used on the Israeli 50 Shekel note. At Bowater Holographics, work by John and Jonathan

Figure 10.30 Colour hologram by Iñaki Beguiristain. Figure courtesy of Iñaki Beguiristain.

Wiltshire led to the completion of a prototype volume hologram replication system, the PolyCopier*, using Bayfol HX photopolymer.

2015 At the age of 89, Graham Saxby, author of three editions of *Practical Holography* [18], died. Martin spoke at Graham's funeral of a man who had inspired a generation of holographers with his kind manner and broad technical expertise.

Bowater Holographics' development of the mass production system for full-colour reflection security photopolymer holograms using the Bayer HX film was commissioned successfully and was recognised by IHMA at the Shanghai Conference 2015.

The Bayer Materials Science group at Leverkusen, which produces the HX photopolymer, became an independent company, Covestro, in September 2015. The company is now legally and economically independent, but will remain a subsidiary of Bayer AG [16].

The International Symposium on Display Holography (ISDH), organised in St Petersburg at the ITMO University, was an opportunity to display the "OptoClone*" holograms which have been made by agreement between the St Petersburg "Faberge" museum, the Hellenic Institute of Holography and ITMO University.

Iñaki Beguiristain [19] produced an image-planed true-colour hologram of truly staggering quality (Figure 10.30). Ironically, this was of a similar subject matter to the ground-breaking work by Paul Hubel in 1988 and demonstrates the incredible developments which have occurred in colour holography with new materials and lasers in the hands of an expert practitioner. Note also that, unlike Hubel's recording on three separate Ilford plates, Iñaki's recording is on a single, multi-chromatic HRT plate. His work in pseudo-colour holography has received worldwide recognition, as mentioned in Chapter 6, but he shows in this "true-colour" piece that he is also capable of world class work in this multiple-laser format.

Notes

* Hologram images from my career at Applied Holographics are reproduced by kind permission of Dr Paul Dunn, Director of Optical Technologies, Opsec Ltd. www.opsecsecurity.com

1 Bjelkhagen, H. and Brotherton-Ratcliffe, D. (2013) *Ultra-Realistic Imaging: Advanced Techniques in Analogue and Digital Colour Holography*. CRC Press (ISBN: 978-1439827994).

2 Hecht, J. (2010) *Beam: The Race to Make the Laser*. Oxford University Press (ISBN: 978-0199738717).

3 Smith, H.M. (1969) *Principles of Holography*. John Wiley & Sons, Inc., New York (ISBN: 0-471-08340-2).

4 Hecht, E. and Zajac, A. (1974) *Optics*. (ISBN: 0 20102835 2).

5 Saxby, G. (1980) *Holograms: How to make and display them*. Focal Press (ISBN: 0-240-51054-2).

6 Visionaries © 2016 Hasbro. Used with permission.

7 Lasermet Ltd. 137 Hankinson Rd, Bournemouth, Dorset BH9 1HR www.lasermet.com

8 Walker, J. L. (1987) "In situ color control for reflection holography", S.M. Thesis, Department of Architecture, Massachusetts Institute of Technology.

9 E. I. Du Pont de Nemours and Company. www.dupont.com/products-and-services/packaging-materials-solutions/anti-counterfeiting-solutions/brands/izon-3d-hologram-technology.html

10 International Hologram Manufacturers Association http://www.ihma.org

11 Bjelkhagen, H. I. (1995) *Silver-Halide Recording Materials: for Holography and their Processing*. Springer (ISBN: 978-3540586197).

12 D. R. Winterbottom, SoundPlan UK&I (David@soundplanuk.co.uk)

13 Richardson, M. (2004) *spaceBomb: Holographics, 1984–2004*. THIS Limited Publications (ISBN: 0-9538-924-0-9).

14 Nike Hologram https://www.youtube.com/watch?v=zNvySWU5AH4

15 van Renesse, R.L. (2004) *Optical Document Security*, 3rd edition. Artech House Publishers (ISBN: 978-1580532587).

16 BMS and Covestro http://www.bayer.com/en/covestro.aspx

17 Klastech Laser GmbH Konrad-Adenauer-Allee 11 44263 Dortmund Germany www.klastech.com

18 Saxby, G. (2003) *Practical Holography*, 3rd edition. CRC Press (ISBN: 978-0750309127).

19 Real colour hologram by Iñaki Beguiristain. www.inaki.co.uk.

Epilogue: An Overview of the Impact of Holography in the World of Imaging

"As I sit here now, I see you in 3D. So why belittle that part of our existence? Why not use it? If everything moves along and there's no major catastrophe, we're headed toward holograms, quite honestly…I would. I don't think there's a subject matter that can't absorb 3D, that can't tolerate the addition of depth."

Martin Scorsese, November 2011.

Figure E.1 My hologram of Martin Scorsese.

When I recorded film director Martin Scorsese in a ruby pulse hologram (Figure E.1) in 2004, it was before the recent resurgence of 3D movie making and, in fact, had nothing to do with any cinematography project he was currently engaged in. Instead, it was part of a project planned to capture and document a selection of creative people that began in the early 1990s, a decade of excess and therefore endless possibilities. During the recording, Scorsese and I talked at length about the general concept of 3D imaging and Project "Time-Slice"; its other participants, writers such as Will Self and artist Sir Peter Blake (Figure E.2), David Bowie and now, Martin Scorsese, the quintessential filmmaker, all recorded as three-dimensional, life-like holograms using a ruby pulse laser. During the recording session, Scorsese quick-fired questions about the future of 3D and joined me in the processing darkroom to continue our conversation. Indeed, such was his enthusiasm that he rolled up his shirt sleeves to help process the glass holograms using photographic chemical formulas from the dawn of photography, in the intimacy of the darkroom. He was absolutely astonished with the result!

The Hologram: Principles and Techniques, First Edition. Martin J. Richardson and John D. Wiltshire.
© 2018 John Wiley and Sons Ltd. Published 2018 by John Wiley & Sons Ltd.
Companion website: www.wiley.com/go/richardson/holograms

Figure E.2 The author's hologram of Sir Peter Blake.

In this unusual case, the exceptional coherence of the laser pulse had captured the entire volumetric space of the room, complete with all its occupants, in ultra-fine detail. His holographic portrait was a great success and he later hung a copy in his New York City office.

I didn't know it at the time, but Scorsese shared an interest with the early films of Louis and Auguste Lumiere, who, to a certain extent, had brought cinematography to the masses. In fact, Georges Méliès's magical dream-like visions in *Le Voyage dans la Lune* were something Scorsese was later to recreate in 3D for his movie *Hugo*, his first venture into 3D moviemaking, thus capturing the essence of Méliès's early technological adventure, whilst advancing Scorsese's skills as a film director to yet another level. Scorsese, then, as one of the world's leading film directors, is a confirmed believer in 3D technology in films [1].

Incidentally, holography appears a natural beneficiary of any escalation in the 3D film industry, since it is able to provide hard-copy imagery for promotional material and souvenirs, for example with companies such as Colour Holographic providing large-format, full-colour multiplex transmission holograms using actual movie sequences. The great movie director Alfred Hitchcock was known for his experimentation with new camera technology, processing of film and post-production editing techniques. During his film career he handled 3D in his own masterful style. His history of enthusiasm for special effects means that I believe it would not be at all surprising to find him working with holography had he still been alive today.

Incorporating suspense into art holograms is very important. "Primal Scream" from my series VORTEX, shown in Figure E.3, reveals my head at floor level, apparently decapitated. My features are confronted by a hand, severed at its wrist, fingers torn to the knuckles. The hand's substance seems as real as the head, which turns in fear from the mutilated hand, fearful of impending doom. The mouth is open, stretching wide to scream, revealing a brilliant light, beaming from my mouth by means of a mirror – a scream, unable to make a sound, it looks like a molten furnace. (I was aware that the technical term used by holographers for the excessive radiance of light from objects inefficiently recorded such as this is, in fact, *noise*, otherwise known in holography as "burn in".)

Developments in production techniques of pulsed holograms and holographic stereograms have combined to provide high-quality three-dimensional illusions that echo the apparently innate need of society to replicate itself through artificial means. A commercial platform has been found for archetypical illusions of this type through the mass production and distribution of embossed stereograms that depict popular celebrities from the music industry.

Figure E.3 Photograph of the author's hologram "Primal Scream".

As hologram recordings of the rich and famous become more common and celebrities queue to join the holographic hall of fame, the author asks:

"Is it documentation, or entertainment, that is shaping the future of holography?"

Where technology is concerned, it can often be very unwise to speculate on the future. Holography is still a relatively young recording medium, much younger than photography, and its theoretical limitations are not yet entirely understood. One of the intentions of this book is to help provide a clear and straightforward route for people wishing to throw their hat into the ring of holography; to enter with a functional understanding of progress so far.

Some of the applications of holography will develop new advances. Several years ago, one may have attempted to record a moving hologram directly using a technique called *multiplexing*. A pulse laser master could (and still can) be made with multiple channels by manually masking a recording plate to allow multiple exposures of a moving or changing subject. Today, however, we are lucky enough to have at our disposal new types of advanced electronics which include plenoptic cameras, whose light-field-capturing capabilities resemble the principle of holography ("*the entire message*"), and which are able to capture instantaneously more spatial information than the conventional camera; and spatial light modulators (SLMs) which offer a new route to the instant transfer of data onto the holographic origination table. There are signs that such digital micro-mirror devices (DMDs) are opening new doors that will advance real-time holography and even lead to holographic television. In the case of television, the huge amount of real-time data input required necessitates a bandwidth far greater than is currently available. However, ground-breaking work being undertaken in this area by Professor Michael Bove and his research team at Massachusetts Institute of Technology (MIT) looks to break through many of the limitations thought to restrict bandwidth and so yield the resolution required for potential holographic broadcast.

In the 1970s when physicist Lloyd Cross introduced his famous "integral stereogram" featuring Pam Brazier, known as "The Kiss", a number of artists including Andy Warhol and surrealist Salvador Dali were "sold the dream" of holographic imaging. Dali's cooperation with Alice Cooper resulted in the image which is still to be seen in the Dali Museum.

The integral holograms developed by Lloyd Cross start life as a conventional film sequence. This film itself portrays no conventional 3D parallax but when the viewer looks through the hemispherical hologram window, each eye looks through a different image slit, just as in a conventional stereoscopic viewer, to see a small variation in angular perception of the subject. Such subjects are encouraged to move during the recording to add the *dimension of time* to the holographic experience. However, if the subject moves too quickly, this results in an abstract form. For example, if the sitter waves a hand faster than the moving camera can accommodate it, the arm may appear to the viewer to bend, and this rather interesting result, unique to this type of holographic image, is referred to as *time smear*.

The aphorism "inside every camera is a hologram waiting to escape" adopts new meaning when applied to holographic stereograms. It's tempting to be puritanical about stereograms; to think of them as a lesser form of holography, as the poor relation. But the fact is that a stereogram, like other time-based media, allows us the luxurious added dimension of time and, thus, motion. A stereogram may be seen as a mechanism of optical illusion, which, like a flicker book, has a beginning and an end. Unlike other forms of stereo photography, however, our cognitive interaction with a stereogram produces the illusion of three-dimensionality and movement without the need for a viewing device, and makes whatever disenthralling knowledge we may have about the stereogram uninfluential over the experience it provides. Just as a celebrity is not a person, it is a construct created for media and the marketplace, a surface appearance that masks the inner reality; the true inner identity of the celebrity figure is different to the perception the public has of them, especially their adoring fans; the 3D stereogram may be an interesting blend of these private and public images. One person profoundly interested in this *ambiguity of reality* was the late David Bowie. In 1999, I was lucky enough to meet him in the Chung King penthouse studio in Manhattan to view and discuss my work, and I found that he, like many others, was immediately intrigued by the realism of holograms, with a particular interest in the future of digital holograms. Weeks after our meeting, he confirmed that enthusiasm with an invitation to film stereograms at the Big Sky Studios in London and asked me to produce a significant order of lenticular stereograms for a special edition of the *Hours* album. This was filmed on a 35 mm Mitchell movie camera with my co-author as I choreographed Bowie's movement for the most dramatic effects.

Mapping consumer demand against technical innovation is by no means straightforward: there is a range of technological, social and regulatory factors at work. But, as with many sectors and innovations, when it comes to thinking about the future, it is often helpful to draw on the ideas and inventions of science fiction. And when it comes to immersive entertainment, *Star Trek* has already provided us with its pre-eminent paradigm, in the form of the "Holodeck", the virtual reality facility on the Starship *Enterprise*. It was here that crew members could immerse themselves on distant planets and interact with aliens without ever leaving their spacecraft.

Since Fox Talbot's early experiments in the nineteenth century and, more relevant perhaps, the early experiments by pioneering photographer Eadweard Muybridge,

Figure E.4 Opto-Clone egg hologram.

photography has represented objects of the surrounding world in a far from perfect way; a two-dimensional vision which has maybe almost reached its conclusion, because it has now reached its limitation, but has already influenced society beyond all expectations.

Holography's history is a continuous evolution involving all kinds of new interests and rarified problems, and it will not be long before holograms of landscapes will be made in formats the size of a cinema screen. Holography began with the straightforward aspiration to produce facsimiles of real objects. This ambition remains today in the "Opto-Clone" technology such as the Denisyuk hologram shown in Figure E.4 [2]. Quality has reached a point where we have no way of knowing whether the photograph replicates the hologram, or the original artefact! For the scientist, this achievement may appear to be the Holy Grail of holography.

But, like photography before it, *digital* holography is certainly more than a recording medium. It is a tool that can also substantiate fantasies that previously never existed, or comment subtly on the real world, blurring our grasp of that reality to the point where objectivity is submerged in a sea of imaginings. Hologram technology has made it possible to create digital holograms from computer-aided designs. Technology which can reproduce, for the first time in 3D, M. C. Escher's 1960 representation of an ascending and descending stairway. This advance opens the door for subjects other than representation of the actuality of our world, creating vistas that could not occupy real space, only holographic space. The extreme reality of modern holograms challenges our understanding of what we mean by "real", yet their essential ambiguity is as unsettling as their verisimilitude is reassuring, in a post-McLuhan age of "virtual reality" experiences.

Digital holography is arguably the most advanced form of visual recording to date. But are holograms merely recordings of objective visual reality, or perhaps creative artefacts capable of expression, interpretation and deception? The same question was possibly asked of photography in the early twentieth century. Dismissed originally by artists as a mechanical recording medium, it fortunately moved into the hands of pioneers such as Bresson and Brandt; a subtle artistic tool capable of the most delicate expression and artful deception. Selective framing of the image, skillfully executed manipulation of objects and lighting of the scene can transform the banal into the dramatic. Post-production techniques were able to add to the deception, as illustrated by Victorian snapshots of fairies and spirits and the "un-existing" of political rivals by Stalin in the Soviet Union; and very similar effects are available to digital holography. People were added and deleted both from photographs and real existence as easily as pressing the delete button in Photoshop, and whereas those early attempts at virtual reality seem clumsy to us in retrospect, the principles are even more relevant to digital holography, where such graphics programs are fundamentally involved.

Can digital media access reality? Well, possibly, and some creators are trying; people such as Steve Grand. In his book *Growing Up With Lucy* [3], he describes his attempts at building an artificially intelligent android, later to become known as Lucy. I guess Steve Grand's work makes us question what it is to be human, and the birth of Lucy has massive implications for us all. In part four of his book, Steve addresses the human spirit:

> "When people ask me questions about this project, they usually aren't really interested in artificial intelligence, per se. To some extent, neither am I. What they are really interested in is themselves and what AI may or may not have to say about the human condition. They worry about how AI might alter their future, or their children's future, and they worry about what the idea of machine intelligence says about the things that they hold dearest, such as consciousness and emotions. And I don't blame them."

In the age of interactive media, our preoccupation with "reality" is an obsession. But what do we mean by real? "Fake real" such as *Big Brother*-style reality television shows and "docudramas", or "Real fake" such as fantasy virtual reality games and *The Sims*. We enter these media, Alice-like, through the digital mirror that is increasingly likely to reflect our own image in the faces of the moving on-screen avatars!

One of the rapidly advancing features of holography is the quest for "real colour". In the same way that photography and television began with "black and white", holography began with monochromatic renditions of coloured subject matter. (As an additive colour system, "black-and-white" is equally as difficult as "full-colour" in holography.) But all that is changing rapidly now!

But what is meant by "real colour"? In the technical chapters, John has pointed out the anomaly of the use of the term "full colour" when applied to a rainbow hologram. What, indeed, could be more fully coloured than a rainbow? Various protagonists of colour holography have used the terms:

- Full colour;
- Real colour;
- True colour.

In photography, one can only approximate the colour appearance of the original scene.

If holographic stereograms involve an intermediate photographic recording, then, to some extent, the same limitation applies to holography. Perception of colour is part of a complex system that involves the human visual physiology as well as the science and technology of making the artificial representation of the original subject. Perhaps the simple term "Colour Holography" is the most logical name.

Holographers such as Lon Moore [4], John Kaufmann [5] and Iñaki Beguiristain [6] have specialised in "pseudo-colour" holography. They arguably have *ultimate control* of the colour of their images. Others, such as T. Kubota, Mike Medora, Yves Gentet, Alkis Lembessis and Hans Bjelkhagen have relied upon the precise reproduction of the reflectivity spectrum of a subject with respect to appropriate lasers. All of these holographers have produced memorable, technically superb *colour holograms.*

One of the great beneficiaries to date of holographic technology is the security industry, whose utilisation of the technique depends upon the difficulty of executing holographic origination and mass production depends upon its problems and limitations.

In the catalogue of the British Museum's exhibition "Fake?", David Lowenthal [7] was quoted, "Technology has simultaneously promoted the skills of forgery and of its detection". There is no better example of this phenomenon than the use of holography as a security device. The price we pay to satisfy the insatiable demands of public appetites for computer games, virtual reality systems, web space and digital holography is a proliferation of counterfeits: simulacra of these self-same digital illusions of reality.

A security hologram attached to the "genuine" copy will often have the word "original" floating in space above the surface. Ironically, this is the use of an *illusion* to guarantee the authenticity of an *inherently illusory reality*, the raison d'être of holography!

Notes

1 http://www.filmschoolrejects.com/news/martin-scorsese-3d-holograms-nadam.php

2 "OptoClone" egg hologram. Photo by M. Richardson; holograms courtesy of Alkis Lembessis, Hellenic Institute of Holography, Greece (HIH).

3 Grand, S. (2004) *Growing Up with Lucy: How to Build an Android in Twenty Easy Steps.* Phoenix (ISBN: 978-0753818053).

4 Moore, L. (1983) "Pseudo-colour reflection holography." In *Proceedings of the International Symposium On Display Holography* (Ed. T. H. Jeong), Lake Forest College, IL 60045, USA, Vol I, pp. 163–169.

5 Kaufman, J.A. (1983) "Previsualisation and Pseudocolour image plane reflection holograms." In *Proceedings of the International Symposium On Display Holography* (Ed. T. H. Jeong), Lake Forest College, IL 60045, USA, Vol. I, pp. 195–207.

6 Iñaki Beguiristain "The Evolution of Pseudo Colour Reflection Holography" for the 7th International Symposium on Display Holography at St Asaph in 2006.

7 Jones, M. (Ed.) (1990) *FAKE? – The Art Of Deception.* University of California Press (ISBN: 978-0520070879).

Index

The Hologram: Principles and Techniques, First Edition. Martin J. Richardson and John D. Wiltshire.
© 2018 John Wiley and Sons Ltd. Published 2018 by John Wiley & Sons Ltd.
Companion website: www.wiley.com/go/richardson/holograms

Printed and bound by CPI Group (UK) Ltd, Croydon, CR0 4YY

16/04/2025

14658382-0001